Teaching and Learning with Cases

Promoting Active Learning in Agriculture, Food and Natural Resource Education

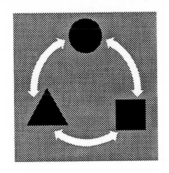

Proceedings of a workshop held in Chaska, Minnesota, July 6-8, 1995.

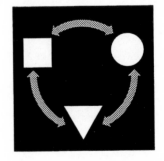

Edited by
Scott M. Swinton

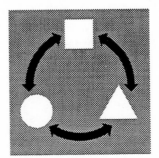

Cooperating Universities:
Michigan State University
University of Minnesota
Oregon State University

Copies of this book may be ordered from the
MSU Bulletin Office, Michigan State University, 10-B
Agriculture Hall, East Lansing, MI 48824-1039,
fax: 517-353-7168.

Price: US $ 14.00, postpaid in the United States.

Michigan State University, University of Minnesota and
Oregon State University are affirmative-action/equal
opportunity institutions.

Printer: Copy World, Inc., Lansing, Michigan

Graphic design: Graphicom, Inc.,
East Lansing, Michigan

First printing: December, 1995

Library of Congress Cataloging-in-Publication Data:

Teaching and Learning with Cases: Promoting Active
   Learning in Agricultural, Food and Natural Resource
   Education / Scott M. Swinton, editor. East Lansing,
   MI: College of Agriculture and Natural Resources,
   Michigan State University, 1995.

ISBN 1-56525-009-5

Table of Contents

Preface v

PART IA. Workshop Proceedings 1

1. Basics of Case Teaching 3

2. Classroom Techniques in Case Teaching 5

   2a. "The Worth of An Oak Tree" Case 10

   2b. Assignment for Session 2 12

3. Team Teaching Experience 13

4. Systematic and Systemic Thinking Tools 18

5. The Basics of Designing Cases into a Course 23

6. Writing and Researching a Decision Case 26

7. Managing the Case Experience 28

8. More on Course Design . . . . : 31

9. Case Use in Extension 41

10. Integrating Decision Case Discussion with Cooperative Learning Strategies 43

11. Case Teaching Issues for Experienced Teachers 46

12. DOTS: A Group-Owned Visual Evaluation Technique 50

13. Creating a Vision for Land Grant Universities in Their Teaching and Learning Programs 53

14. Maintaining the Momentum 55

Part IB: Workshop Cases 57

1. The Worth of a Sparrow 59

2. An Honest Face 87

3. Tom and Joan Karen 91

4. Betsy's Tempeh 96

5. Minto-Brown Island Park: Farming the Urban-Agricultural Interface 109

6. The Case of the Lecture that Wasn't 119

7. Thea Grossman 123

PART II: New Decision Cases 129

Preface to Part II 130

1. When the Cows Come Home 131

2. Quality and Coordination in the Food System: The Case of Riceland Foods 140

3. After the Fire 144

4. Is a Drink of Water a Drink of Weed Killer? 158

5. Black-Mason Farm: What Now? 162

6. Thistledown Farms 169

7. The Boundaries of Chemical Drift 185

Appendices 193

A1. Abstracts from the Case Showcase 195

A2. Case Plans Written in the "Researching and Writing Cases" Session 200

A3. Agenda for the Workshop 217

A4. Workshop Facilitator List 221

A5. Participant Roster 224

A6. Selected References on Decision Cases 234

A7. Suggestions for Maintaining the Momentum 235

Preface

As its title indicates, the workshop "Teaching and Learning with Cases" centered on *active learning processes* using decision cases. It was not a conference at which attentive audiences listened to papers being lectured. Rather, it was an interactive workshop primarily organized into parallel, participatory sessions.

The workshop opened with a welcome by organizers Steve Simmons, Eunice Foster, and Anita Azarenko, on behalf of the 15-person planning committee from University of Minnesota, Michigan State University, and Oregon State University. It moved immediately into the case method with a plenary session at which Kent Crookston and Joe Broder led case discussion of "Worth of a Sparrow" and "An Honest Face." Both cases raised thorny ethical issues.

At the heart of the workshop were three decision cases taught by workshop participants to one another. Participants were divided into four groups of roughly 20 each for parallel case sessions. The groups were divided into three teams, each responsible for teaching one of the three cases. After discussing techniques in case teaching, participants planned the teaching of their cases on the evening of Thursday, July 6. On Friday, they taught the cases, each subgroup acting as teachers for one case and learners for the other two.

Concurrent group sessions later Friday and on Saturday offered a smorgasbord of opportunities to learn about special topics in case teaching and learning. These included how to design cases into a course, how to write cases, how to manage the case experience in class, how to integrate cases into cooperative learning strategies, and how cases can be used in Extension outreach programs. In the session on writing and researching decision cases, workshop participants drafted the 47 decision case planning outlines in Appendix A2. Other ideas were displayed Friday evening at the Case Showcase; these exhibits are abstracted in Appendix A1. The workshop closed on a forward-looking note with a talk by Tom Warner of the W.K. Kellogg Foundation on the future of land-grant university teaching and a final session on how to maintain the case teaching momentum acquired at the workshop. Details on the sequence of the workshop can be found in the agenda reprinted in Appendix A3.

The experiential, interpersonal nature of the workshop presented a challenge to recording the proceedings on paper. In preparing these proceedings, we have relied on notes taken by recorders and session facilitators. Our editing has aimed to preserve the spontaneity of the sessions, while attempting to eliminate redundancies. All decision cases used at the workshop are reprinted here.

"Something Old, Something New..."

We have included in the proceedings the eight decision cases used at the workshop; "The Worth of an Oak Tree" appears in Part IA with the session on classroom techniques, while the rest appear in Part IB. Except for the one case used as a reference ("The Case of the Lecture that Wasn't"), all were taught at the workshop. Part II adds a set of cases not previously published. Teaching notes for these cases are available from their authors. Each case was reviewed by at least two peers who are experienced in case teaching, most of them participants in the workshop. Details are provided in the opening of Part II.

Background and Acknowledgments

The workshop "Teaching and Learning with Cases" was a collaborative endeavor from start to finish. During the year from planning to execution in Chaska, Minnesota, July 6-8, 1995, the workshop involved complex interactions among many people. It was sponsored by three universities, planned by 15 people, and experienced by 90 participants (listed in Appendix A5). Several of the workshop planners had participated in the University of Minnesota's October 1991 conference in Chaska on "Decision Cases in Agriculture." Others were involved in a regional workshop on "Teaching with Decision Cases" held at Michigan State University in August, 1994. At that meeting, representatives from Michigan State University, the University of Minnesota and Oregon State University began to discuss the possibility of a national workshop focused on teaching and learning with cases. The 1995 workshop grew out of shared interests in 1) moving beyond case research to case teaching, and 2) expanding the subject matter scope beyond production agriculture to embrace food and natural resources.

The conference facilitators, who are listed in Appendix A4, spent innumerable hours planning the workshop sessions and discussing desired learning outcomes in more than a dozen telephone conference calls and countless electronic mail messages. In fact, many of the facilitators met in person for the first time in Chaska, the day the workshop began. Workshop logistics were skillfully managed by Tammy Dunrud, coordinator of the Program for Decision Cases at University of Minnesota, and Lisa Brienzo, conference coordinator.

For these proceedings, Ray William of Oregon State provided tremendous help in planning and serving as outside review coordinator for the "After the Fire" case. Ray, Anita Azarenko, and Eunice Foster all served as proceedings recorders in many of the workshop sessions. Ken Fettig did painstaking copy-editing of the workshop proceedings and Ken Dettmer designed the cover and layout. Patricia Neumann typed many revisions with great patience and good humor.

Financial support for the workshop was provided directly by the W. K. Kellogg Foundation and indirectly by the U.S. Department of Agriculture (USDA). In late 1994, as conference plans moved forward, the Kellogg Foundation was already providing support to the University of Minnesota and Oregon State University through its Food Systems Professions Education (FSPE) project. Kellogg generously agreed to support this workshop in furtherance of those goals. USDA had awarded a Higher Education Challenge Grant to Michigan State for developing new cases and improving case-based pedagogy in agriculture and natural resources. We thank both organizations for their contributions to making possible the national workshop on "Teaching and Learning with Cases."

took place in July, 1995, the University of Minnesota has established an Internet list server called CASENET. CASENET is developing into both a medium for discussing case teaching experiences and a means to keep abreast of new case publication and teaching developments. To subscribe, send an electronic mail message to Tammy Dunrud at: dunru001@gold.tc.umn.edu. Once on the list server, the CASENET address is: casenet@maroon.tc.umn.edu.

Apart from the selected references in Appendix A6, further information on cases in agriculture, food, and natural resources is available from two sources. The Program for Decision Cases at the University of Minnesota maintains a library of cases that can be used for a small fee. The cases are described in a case catalog published by the Program. The catalog and other information are available from:

Ms. Tammy Dunrud
Program for Decision Cases
College of Agriculture
University of Minnesota
1991 Buford Circle
411 Borlaug Hall
St. Paul, MN 55108

A new series of decision cases is currently being developed by the Michigan State University College of Agriculture and Natural Resources Office of Academic Affairs. Details on these cases as well as additional copies of this proceedings can be obtained from:

MSU Bulletin Office
10-B Agriculture Hall
Michigan State University
East Lansing, MI 48824-1039

Fax: 517-353-7168

Scott M. Swinton
East Lansing, Michigan
November, 1995

Epilogue

Participants made many thoughtful suggestions on how to maintain the momentum developed in teaching with cases in agriculture, food, and natural resources (Appendix A7). Since the workshop

Part IA
Workshop
Proceedings

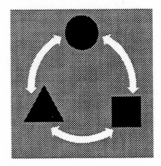

The Basics of Case Teaching

Presenters: *Josef Broder, University of Georgia, and Kent Crookston, University of Minnesota*

To familiarize workshop participants with the basics of case teaching, the assembled group was plunged into discussion of two introductory cases. These cases had been chosen specifically for an audience of scientists and educators, and participants had been asked to read them ahead of time. Each case discussion was led in socratic fashion by its author—"The Worth of a Sparrow" by Kent Crookston and "An Honest Face" by Joe Broder. The cases were taught one after the other, with no opportunity for debriefing. Debriefing and participant reactions began in the following session on "Techniques in Case Teaching." These two cases, as well as the others used at the workshop, appear in Part IB. Below, the authors describe the cases and what they hoped to elicit from the case discussion.

"The Worth of a Sparrow"

"The Worth of a Sparrow" was developed as a general interest case to engage anyone working in the life sciences. The principal objective of the case is to provide users an opportunity to wrestle with the troublesome issue of "rights" of interest groups which disagree with methodologies employed by researchers. A less obvious, yet more important, objective is to help users understand the potential for, and to work toward, synergy from a background of differences. Specific sub-objectives include:

a. to have students evaluate merits of the University of Minnesota's bird trapping program, and to defend a recommendation for continuation or termination of the practice.

b. to consider whether and how to research the complex phenomenon of bird damage.

c. to recommend whether the university continue on-campus cereal research or transfer the programs to rural locations.

The case was first taught as the opening case for the 1991 national workshop on "Use of Decision Cases in Agriculture" sponsored by the University of Minnesota in Chaska, Minnesota. Since then, the "Sparrow" case has been taught many times, and we have noted that it is most effective with audiences that are not dominated by scientists or educators. A strong divergence of opinion within the group is most helpful. The challenge in teaching this case is to focus the discussion on a synergistic resolution, rather than mere compromise.

"An Honest Face"

"An Honest Face" was designed to provide a forum for discussing academic dishonesty in higher education and was written for workshops and seminars on faculty development. The case was first presented at the Twelfth Annual Conference of the World Association for Case Method Research and Application, Leysin, Switzerland, June 18-21, 1995 (Broder).

The case examines the personal nature of academic dishonesty and how instructors and institutions develop policies and techniques for protecting the academic integrity of the classroom. The case can be presented to faculty, instructors, and graduate student audiences where the topics of discussion are teaching techniques, academic integrity, and professional development.

A popular young teacher encounters his first serious instance of possible cheating in the classroom. Lacking clear and convincing evidence of the cheating, he searches for a solution to the problem. The challenge faced by the teacher is twofold. First, he must deal with his own feelings of anger and betrayal. Second, he must protect and defend the integrity of the university in dealing with matters of academic dishonesty.

The case opens with Professor Jenkins compiling final grades for the course. He discovers grade irregularities and suspects that some students may have cheated on the exam. Most troubling is that one of the suspects is a student he had grown fond of during the quarter. The timing of the incident creates additional pressure for a swift solution. Given that the suspected cheating occurred on the final exam and that many students had left campus for the summer, possible solutions to the case are severely limited.

The case describes the events that led to the suspected cheating. The audience is asked to evaluate the people and circumstances that may have contributed to the problem. The case at hand has both personal and institutional dimensions. At the personal level, the issue is one of dealing emotionally with betrayal and disrespect. At the institutional level, the issues are maintaining the integrity of the academic process. The audience is not provided with a pre-packaged solution to the problem, but rather is asked to explore the sources of the problem and identify solutions. The case should be familiar to those who have contemplated cheating or have been the victim of cheating. Lest we judge the students in the case too harshly, consider the expression, "virtue is the absence of temptation."

References

Broder, Josef M. "An Honest Face: Academic Dishonesty in Higher Education." in Hans E. Klein, ed., *Teaching and Interactive Methods*. Needham, MA: World Association for Case Method Research and Application, 1995. Pages 339-348.

Classroom Techniques in Case Teaching

Group B Report

*Facilitators: William M. Welty, Pace University;
Josef M. Broder, University of
Georgia*

One objective of the conference on "Teaching and Learning with Cases" was to give participants a wide range of experience with case teaching. Following an orientation session, conference participants played the role of students in two separate case presentations, "The Worth of a Sparrow" and "An Honest Face." Immediately following the case presentations, the participants were divided into four separate groups and break-out sessions were held to analyze and discuss the case presentations. The following summarizes the discussion by Group B.

Participating in the Case Discussion vs. Discussing the Case Presentation

In case teaching, a key distinction is made between participating in the case discussion and discussing the case presentation. During the actual case presentation, participants are expected to play the role of students in resolving the problem/situation presented in the case. Likewise, the facilitator guides the discussion in the context of the case. The roles played by participants and facilitators during the case presentation are analogous to the characters in a drama. They may be aware that the session is staged, but there is a need to pretend or role-play the situation. Questions about the case method are deferred until the end of the case presentation. Ordinarily, participants are asked to delay comments/questions about case teaching until the end of the session. Some case teachers may temporarily suspend the case discussion, analyze events that have occurred in the presentation, and continue the case presentation. While this technique may help participants see specific points about the presentation, the continuity and flow of the case suffers (much like commercials affect network television shows). For this conference, the analysis of the presentation was scheduled for the break-out sessions.

The process of analyzing and discussing an actual case presentation is known as debriefing the case. This exercise encourages participants to discuss their impressions and ask questions about the manner in which the case was presented and discussed. At the beginning of the break-out session, participants were asked to take a few minutes to reflect on the previous case presentations. Then, they were asked to write down their answers to the following questions:

1. What were your feelings during the case presentation?

2. What questions do you have about how the case was presented?

Next, the facilitator of the session asked participants how they responded to the questions and recorded them on the chalkboard or flip chart. Responses to the first question included feelings of frustration, excitement, anger, confusion, enthusiasm, impatience, and others. The lesson to be learned from this question is that case teaching is an emotional or affective experience, in addition to being an intellectual or cognitive experience. Advocates of the case method argue that feelings are an important part of learning and retention; students learn more when the case evokes an emotional response, and these emotions motivate students to participate in the case discussion. Anecdotal evidence suggests that students tend to remember cases for longer periods and with greater clarity than content learned in more passive teaching environments.

Participant Frustration

Facilitators should be sensitive to the level of frustration felt by the participants. These frustrations can stem from the case or the manner/setting in which the case is facilitated. Cases with no clear courses of action are intended to create a certain amount of frustration among participants. These feelings stimulate discussion and a desire to identify causes and solutions to the case situation.

Sources of frustration were identified for "Worth of a Sparrow" and "An Honest Face." The group indicated that the room layout and the size of the group were not well-suited for the case presentation. The auditorium-type seating arrangement made dialogue among participants difficult; lines of sight were poor; and the acoustics of the room were lacking. Some felt that the group was too large for the time period in which the case was discussed. The group was frustrated with the flow and direction of the discussion. Some complained that the discussion did not lead to decisive action.

The facilitator's response to these criticisms was that the opening case was designed to create a certain amount of frustration or curiosity in the case method. The large setting and group size served to illustrate how the room environment affects case discussion. At the outset, the conference organizers agreed that the facilitators would refrain from writing notes on the chalkboard during the case discussion. In retrospect, we believe that this may have increased the level of frustration in the presentation. That is, an important tool for acknowledging comments and giving direction to the discussion was purposely omitted from the opening cases. In all candor, the conference organizers did not fully anticipate how the group would respond to the opening cases and were somewhat surprised by the outcomes.

Responses to the second question can help students analyze how the case was facilitated. Some generic questions might include asking facilitators how they decided on an opening question; if the case discussion had met their expectations; or why they probed certain questions/comments and not others. These are generally safe questions and should be answered to the best of the facilitator's ability. Question No. 2 can generate questions that may not warrant a response. For example, some participants might ask, "Tell us what really happened," or "What action do you think should have been taken?" Facilitators should avoid answering these questions for fear of confirming or negating comments and suggestions made during the case. The facilitator's opinion of the case or knowledge of events not mentioned in it are irrelevant and should not be imposed on the case discussion. When asked such questions, the facilitator can redirect the question back to the student or to another student.

Benefits of Debriefing Sessions

Debriefing sessions can also benefit facilitators and case writers. The quality of a case cannot be fully known until it is presented in an actual case discussion. Stories that read well on paper may not be effective in case discussions. Hence, case writing and presenting becomes an iterative process, whereby the case is edited for more effective discussion. The facilitator can use the debriefing information to tweak the case, to analyze the opening questions, the probing and redirection of questions; the shifts in perspective, closure, etc. Also, the facilitator can use the debriefing to edit his/her user's manual or teaching notes. Some case teachers/writers modify their teaching notes after every presentation. However, such practices may inadvertently add too much structure and rigidity to the case, leaving little room for uniqueness and creativity in case discussions. Keep in mind that the facilitator's role is to encourage discussion and dialogue and not necessarily to lead the discussion down a well-rehearsed and well-traveled path. Actions recommended by one group at one point in time may be entirely different from those recommended by another group in another setting. One of the important ironies of the case method is that we are concerned as much with the process whereby actions are recommended as we are with the actions themselves.

Group C Report

Facilitator: Chris Peterson, Michigan State University

Each person in the room was asked to introduce himself or herself by providing some personal background and "one special thing about yourself."

Participants were than asked to write answers to the following:

1. What were your feelings during the case discussions in the prior session?
2. What questions do you have about the process?

(Peterson elicited responses from the center of U-shaped tables and posted participants' comments on two flip charts.)

Tally of answers:

Feelings about Case Discussion

- Lost
- Frustrated
  - no chance to speak
  - no closure
- Openness/sharing
- Intimidated
- What is the value of it? What are the goals, objectives?
- Involved and surprised to be

Questions about the Case-Teaching Process

- Was there purpose in technique?
- Where does problem-solving happen?
- What should the learner be doing?
- Is resolution an objective?
- Minimum amount of information in a case?
- How to control the time?
- How to get participation?
- How to control the process?
- What is case objective in course?
- What are options? Who should know?
- How to get closure?
- How to grade participation?
- How to develop skills and knowledge?

Practice Teaching a Mini-Case

Peterson elicited comments and controlled flow of discussion by posting comments on flip charts. He noted the contrast between problem cases and opportunity cases. He also noted that he could have "lectured" to the participants about how students feel when thrown into the case method or about the list of key questions. But the power of the method is that the participants now know the feelings students have and that they have generated the list of key questions for themselves. He asked participants to wait for the answers to their process questions until after they have worked with teaching a case.

The assignment of "Worth of an Oak Tree" was then made. Participants were divided into five teams, given 30 minutes to read the case and plan the first question they would use to open the discussion of the case.

Proposed Opening Questions

Group 1

Question: How should the judge decide?

Teaching Objective: What is the value of public vs. private property?

Group 2

Question: Who has the *right* to the *property*?

Teaching Objective: Focus on key, italic words.

Potential Audience: Content for rural appraisers, financial analysts and process skills for educators.

Group 3

Question: If you were the judge, would you condemn this property?

Teaching Objective: Develop a role play focused on the judge.

Group 4

Question: What should the county commission do once the judge renders his decision?

Teaching Objective: Discuss possible outcomes from the judge's decision and for the tree's survival; focus on value of tree and what determines it.

Group 5

Question: Who are involved in this issue and what are their objectives?

Teaching Objective: Analyze roles; leave flexibility of focus for different class types.

Peterson summarized two general approaches to starting a case:

a. Analyze the situation first and then move to options.

b. List options first and move backward into analysis.

A question was raised concerning student attempts to preempt the discussion with some less valuable issues or approach. Peterson responded that the case teacher should avoid the student's points rather than confront them directly. Confrontation tends to reduce future participation because many students become fearful of the process.

Assigning Cases to Each Team

The participants were then divided into their three teaching teams and the case assignment made to each team by lot. The teaching teams were then asked to prepare the following for Friday's teaching sessions:

1. Prepare to teach the group's assigned decision case to the other two groups.
2. Prepare to be students for the two cases being taught by the other groups.
3. Prepare study questions for the students of the case being taught, and post the questions on the board by 9:30 p.m. that evening.

Peterson then gave some limited guidance for how the groups should proceed.

- Use "Worth of an Oak Tree" as a guide to what you must do.
- Think about how creative you want to be.
- Think about how much sharing of discussion leadership you want within the group.
- As a group, do you want to give preference to the least experienced in the group?
- Avoid leading class to instructor's pre-determined "right answer."

Group A Report

Facilitators: Steve Simmons, University of Minnesota; Andy Skidmore, Michigan State University

Overview

The session began by debriefing the two sample cases experienced by the participants in the previous session on "The Basics of Case Teaching." This discussion was followed by a small group exercise (see "Group B Report") using a model case "The Worth of an Oak Tree." This exercise was designed to acquaint the participants with the choices faced by a case teacher in deciding how to conduct a case discussion. In the final phase of the session, case teaching teams were identified and assigned a case to prepare to teach the following day.

Debriefing of the Opening Two Cases

We opened discussion of "The Worth of a Sparrow" and "An Honest Face" by asking participants to express how they had felt during the two model case discussions. Feelings expressed ranged from "frustrated" to "engaged." Some were frustrated in a positive way by being pushed out of their comfort zones and learning to appreciate other people's biases. Others were frustrated because they couldn't recognize any clear objective or terminal point for the discussion. They felt they were "left hanging" at the end. Some felt perplexed at the lack of structure, as the discussion constantly vacillated from one side of an issue to another. We concluded that the sense of frustration originated chiefly from confusion regarding the purpose of the cases. The facilitators noted that this was due in part to the fact that these cases had not been approached out of any context other than to have the participants experience a case discussion. On the positive side, participants also felt affirmation of their perceptions. ("Hey, someone else agrees with my views.") Many questions were raised about the process of teaching with cases. Specific questions to be answered during the rest of the session included:

A. Is there a best process or right way to teach with cases?
B. Should a case be tied to a theoretical framework?
C. Should discussion be structured or free-flowing?
D. How do you control discussion?
E. How to assess if students have read the case before class?
F. How to increase student participation?

Preparing to Lead a Case Discussion

Simmons opened the case preparation exercise by describing case teaching as a three step process. The first step is *Assignment and Engagement.* Students are usually given an assignment before the discussion. This assignment could range from something as simple as reading the case to a more complex task such as evaluating case information or doing background reading. The teacher might also use audio-visuals to enrich the students' understanding of the case. This step gets students' involved and feeling some sense of ownership in the case. The second step is *Analysis and Discussion.* The students systematically analyze the case and discuss it under the direction of the teaching objectives. The third step is for the students to *Respond* to the case discussion and analysis. A decision does not always need to be made, but one usually is. Response can take many forms, such as recommending what decision should be made,

expressing feelings during case discussion, conducting further case analysis, or identifying additional information needed to make an informed decision.

The exercise for preparing to lead a case discussion (see "Group B Report") was based on the two-page case "The Worth of an Oak Tree." Participants proposed a wide range of opening questions. Some openers were very global, such as "What's going on here?" or "Is there a problem?" At the opposite extreme, other opening questions volunteered were as specific as "Under what circumstances does the government have the right to take private property?" "Who owns the oak tree?" "How much would you donate to save the tree?" "Is there a win-win solution?" or "If you were the judge, what decision would you make?" Some questions introduced a hypothetical additional character to the case (e.g., "As the mayor, what would you do to heal the community?").

Global versus Specific Questions

Each type of opening question has its merits. A global question can be used to get all the issues out; it doesn't limit the direction of the discussion, and it helps students identify problems. A specific question immediately focuses discussion on the subject of the question. For less mature students who tend to focus immediately on the solution, a specific opening question going right to solutions will engage them and spark their interest. When specific questions are used as discussion openers, the teacher must be prepared to change the direction of the discussion if the question is answered with a short definitive answer and doesn't prompt further discussion. Introducing an additional or hypothetical character as the decision maker in an opening question can serve to broaden the case's perspective.

A typical case discussion includes questions designed to address at least four kinds of outcomes: 1) definition and clarification of facts and stakeholders, 2) definition and analysis of key issues, 3) definition and analysis of decision options, and 4) elicitation of a final decision response. We explored how the opening questions contributed by each of the small groups accomplished one or more of these outcomes.

Dave Hamby / Associated Press

The Worth of an Oak Tree

Introduction

In 1990, an oak tree growing in a field near Magnolia Springs, Alabama, started a debate between conservationists and property rights advocates. The 500-year-old oak had a trunk that was 25 feet around and branches that extended out 150 feet (half the length of a football field). The root system of the tree was estimated to cover almost an acre of land.

In October of 1990, someone used a chain saw to cut a ring around the trunk of the tree, a process known as 'girdling'. The purpose of girdling was to cut off the flow of nutrients between the roots and the top of the tree, which would lead to the tree's death. Although the damage to the tree was investigated by local police, no suspects were charged.

This case was developed by Steve R. Simmons and Tammy Dunrud, Department of Agronomy and Plant Genetics, University of Minnesota, as a basis for class discussion. Copies may be obtained from Dr. Simmons and the College of Agriculture Program for Decision Cases. Text copyright, 1994. All rights reserved.

UNIVERSITY OF MINNESOTA

Background

For several years, Baldwin County had been trying to buy the 2.8 acres of land on which the long-lived oak was located in order to develop a park. These efforts were not successful because the landowner did not feel that enough money was being offered for the land. Finally, in the summer of 1990, the County Committee decided to force the sale of the land through condemnation[1] because they were concerned that the landowner would not safeguard the tree, which was endangered by new highway construction nearby. The County Committee feared that the highway project would damage part of the tree's root system and threaten its life.

The landowner at the time, 72-year-old Mildred Casey, argued that her ownership of the land included ownership of the tree, which meant that she could do with it as she wished (in other words, exercise her property owner rights). She started a process in court to stop the county's condemnation of the land. She then sold the land to an Alabama businessman, Kenneth Arnold, who continued the court fight over the land begun by Ms. Casey. Mr. Arnold also believed that the land was worth more than the county was willing to pay and stated that he would fight the condemnation "all the way to the Supreme Court."

Tree rescue effort

While the court battle continued, a forester from Florida, Stan Revis, began an effort to save the damaged tree. Soon after the chain saw incident, Mr. Revis spent his vacation at the tree site working to apply bark "grafts" to the trunk to bridge the girdled area and allow nutrients to move between the roots and leaves. A 'Save the Tree Committee' was also started, which set up an air-conditioned and heated tent (nicknamed the "ICU" or intensive care unit) around the tree (see photo). "What we've tried to do is get the tree to heal prior to the stress of summer heat," commented one Save the Tree Committee volunteer. The tent contained an electrically-operated heater and air conditioner to maintain a temperature of 78 degrees, regardless of the weather outside. The Committee also included installing an irrigation system to keep the humidity inside the tent near 100 percent, which helped prevent the grafts and wounds from drying out. Over the first few months after the damage was discovered, an estimated 30,000 people visited the tree and about $30,000 was donated to finance the tree rescue effort.

The decision

In March, 1991, a court judge was to announce his verdict regarding the condemnation by the county of the property containing the long-lived oak.

---

[1] Condemnation is a legal procedure that permits governments to declare property forfeited or appropriated for public use.

Teaching and Learning with Cases
Oak Ridge Conference Center
July 6, 1995
Facilitators: Oran Hesterman, Chris Peterson, Steve Simmons and William Welty

Session 2
Classroom Techniques in Case Teaching

Group exercise

In preparing to lead a discussion of a decision case in a class, a teacher must face a number of decisions and questions. This small group exercise is intended to permit you to consider some of these as you decide how you might teach the decision case titled *The Worth of an Oak Tree.* Your specific assignment is described below. Since you will have only a limited amount of time for this exercise try to avoid spending excessive time on any one point and attempt to move through each question.

1. Read the case individually.

2. Discuss with your group what the case is about and the important issues raised by the case. Why would these issues be important to consider?

3. Discuss and define with your group what your educational objectives would be in teaching this case.

4. Describe how you would assign the case and what approach you would use to engage the students in the case issues before initiating a discussion of the case.

5. Decide on a general strategy that you would use for conducting a discussion of this case. Identify a specific opening question that you would use to begin the case discussion. What would you hope to achieve by using this question?

6. How would you conclude the case discussion? What kind of assignment or other response might you make for your students following the case discussion?

Team Teaching Experience

Group A

Facilitators: Steve Simmons, University of Minnesota; Andy Skidmore, Michigan State University

The team teaching assignment divided participants into three subgroups, each assigned to teach one of the cases: "Betsy's Tempeh," "Tom and Joan Karen," and "Minto-Brown Island Park: Farming the Urban-Agricultural Interface." Members of the two teams not involved in teaching served as students for the team involved in teaching each case.

The three teams approached their teaching assignments in very different ways. Most involved more than one team member in teaching the case, at least as a "scribe" to post student comments to a flip chart. The team or multiple teacher approach worked well but tended to be fairly structured. Each teacher, in turn, had to begin at a more superficial level. Some of the teams clearly modified their approach after seeing and hearing the debriefing of the previous group(s). Issues discussed included alternative questioning strategies, how to facilitate discussion with quiet versus talkative students, approaches for having students respond to a case, maintaining the flow and energy within a case discussion, and use of the blackboard.

Posting Comments: Teacher or Scribe?

Posting comments to a flip chart or blackboard helped to focus discussion. But the task of writing down comments can distract the discussion leader. Having another team member or student act as a scribe allowed the discussion leader to concentrate on the discussion and what was happening in the room. But the scribe could get lost if there was poor communication with the discussion leader. Also, if the leader did the writing, the "down time" while recording comments gave students time to think. It also gave the teacher greater control of class discussion. However, writing on a board tends to force the teacher to turn his or her back to the students, sometimes breaking eye contact. Anyone who is scribing must understand what is happening in the discussion in order to serve as a filter to interpret student remarks. A teacher can use a student scribe as a way to ascertain the understanding or comprehension of the students. It is important that any scribe write down the response of all students. If a student's comments are not recorded, that student may feel that what he or she has to contribute is not important and will be reluctant to become engaged in the case again. If flip charts are used, post them on the wall for future reference.

Controlling Students and the Discussion

To control a dominating or disruptive student, have that person serve as a scribe. Or, the leader can move close to the student and then move away when he or she wants the student to participate again. Most students tend to be quiet when the teacher is close by.

Several strategies can help to control the discussion when a student wants to move it to another major issue and the instructor prefers not to. Record the student's response on a flip chart or blackboard with the comment, "We'll come back to that." Or, ask the student to hold his or her thought. It is imperative that the discussion return to those deferred comments or the student will feel betrayed.

Case teachers can make several mistakes. Both too much or too little structure can stifle discussion or learning. Reading a case aloud after the students have been assigned to read it may make it very difficult to get the students engaged in the discussion or may make them reluctant to prepare for the next case.

Group C

**Facilitators: Chris Peterson, Michigan State
University; Jack Stang, Oregon
State University**

Case: "Tom and Joan Karen"

The prior evening, the teaching group had given
the following assignment:

- Read the decision case.
- Develop a time line of events.
- List two or three concerns of Tom and/or Joan.

To begin the discussion, Nick Jordan gave an
overview of the teaching game plan:

- Plenary discussion of case facts and
  chronology (Nick)
- Small group discussions to identify concerns
  and alternatives for the Karens
- Plenary to list concerns (Dave)
- Plenary to list alternatives (Ed)
- Individuals to write three sentences on what to
  do as the Extension agent working with the
  Karens
- Plenary on what to do (Dave)
- Closing (Nick)

The small groups identified the following at the
end of their individual discussion times:

Concerns

- Farm Bill 1995: Will CRP be continued?
- Depend on Farm Bill
- Self-sufficiency
- Grief over near-bankruptcy
- Future finances
- Brent as a manager
- Property taxes
- Tom's time trucking
- Aging and health
- Opportunity to farm for Brent and Sally
- Loss of quality of life
- Financing for Brent to farm

Alternatives

- Brent should practice farming
- Enroll in goal-setting workshop
- Return to farming
- Sell farm, move to town
- Sell farm, stay in country
- Get family-style truck and keep trucking
- Put farm in land trust, not CRP
- Hire truck drivers
- Tom back to sales
- Change crops

Debriefing on "Tom and Joan Karen"

The debriefing began by asking the teaching
group about their evaluation. Many ideas, both pro
and con, were discussed. Peterson asked the
students if they were disappointed that the class
did not discuss or rebuild the time line as the
assignment had called for. Some were disappointed, but others noted that the professor could
collect and mark the time lines even if not used
directly in discussion.

The students liked the pacing provided by the
pre-announced agenda for the class. They found
that it worked well with this group and was not too
forced. The small group discussions also went well.
They helped increase participation by "greasing the
wheels." The mixed format of the discussion also
enhanced the level of mental involvement.

The discussion had been vague about the nature
of the problems being addressed. In the end, what
connection existed between concerns and alternatives? The context for how the case could create
value for students needed to be better defined. One
farm management teacher said he would not use
this in an introductory course since there were no
exhibits to give the underlying financial information
critical to the discussion. Peterson noted that cases
can be used to introduce and motivate future course
content as well as to apply known content.

Leave Some Mystery

A question was raised: When should the instructor reveal the educational objectives of a case to the
students? Peterson indicated that being explicit
about objectives is not necessarily helpful. The
power of the method is often greatest when there is

some mystery for the students to confront. It is better to err on the side of revealing objectives or means to improve discussion at the end, not the beginning, of the discussion. One effective way to close a case is to have the students discuss how well they performed in discussion.

One of the teaching team asked, "How else could we have used the case? We struggled over how to do it." The teaching options were limited by the lack of exhibits in the case, and the group made a very good choice. One participant noted that the conference setting was not entirely realistic because the cases were assigned and then an objective had to be developed. In reality, the opposite occurs—the objective is set and then a case is developed or found to fit the objective. Cases should be written with the educational objectives in mind, including expected analysis approach and necessary data requirements.

Peterson commended the group for its mix of teaching and learning styles—whole group, small group and individual activity. The leader was always in control, but it didn't feel that way.

Case: "Minto-Brown Island Park"

The prior evening, the teaching group had given the following assignment:

- Read and study the decision case.
- Identify, individually, two creative options for Mr. Anderson-Wyckoff to move toward conflict resolution.

The teaching-team leader assumed the role of the key case decision maker, Anderson-Wyckoff. She welcomed everyone to the commission meeting. Individual students began to take on the roles of various key commission members. They assumed these roles without any instruction from the teacher. The discussion became more and more lively. Little resolution of the problem was ever achieved, but the dynamics of a group process linked to a controversial issue became very apparent to all participants.

Debriefing on "Minto-Brown"

The teaching group itself was concerned that the pace had been slow at the beginning and that more structure may have been needed. For example, the team might have started in the classroom setting by

describing the role play and assigning roles. Their teaching approach was prompted by a desire for the students to experience conflict resolution firsthand. The method did seem to do precisely that.

The students had several reactions:

- Too little structure for group.
- Some students did not understand their role as advisory board members — but they did play roles.
- Assignment should have been to take a position.
- Did we need another facilitator to liberate the "Anderson-Wyckoff" character to be himself?

Stang noted that the case teacher can use role play at the beginning and then change to a different classroom mode later in the hour. Peterson commented that it would be important to debrief the students at the end of the role play. The fact that students did not move toward conflict resolution provided a "teachable moment." The unstructured discussion lead instead to roles becoming rigid and conflict being heightened.

Case teachers should also recognize that a time-out can be taken anywhere in a discussion to debrief the process and then return to the case. The teacher can use the time-out to present a mini-lecture on a key point. For example, a time-out could have been used to present a conflict resolution typology and then ask the class to think about moving toward resolution when the role play resumed. Paul Gessaman, one of the participants, suggested such a typology:

Facts Converge Issues Converge	**Facts Converge** *Issues Diverge*
Facts Diverge Issues Converge	*Facts Diverge* *Issues Diverge*

Conflict resulted in class remaining in lower right cell.

Assigning Roles to Students

Peterson concluded the debriefing with several additional comments. The group chose a high-risk strategy. The discussion leader never played the role of the faculty member. This is a high ambiguity situation for students and also gives the floor to the best student actors in the class. On the positive side, the room was transformed into the meeting room and real emotions and arguments were in evidence. On the negative side, shy students were closed out of the discussion. This risk could be minimized by assigning roles ahead of time. It can be especially effective if students are assigned roles against type, e.g., a farm student playing an Audubon member.

The teacher might even plant a limited number of students who have been assigned roles and not assign roles to the other students. This can backfire if other students realize the discussion has been rigged. One participant suggested that role-play guidelines should be given ahead of time. Students could then be asked to stay in roles and minimize "personal" attacks. Another technique is to have students prepare the case in advance and then make small group assignments to develop roles during the first 10 minutes of class.

Case: "Betsy's Tempeh"

The prior evening, the teaching group had given the following assignment:

- Read the case.
- Is this a success story? Why?

The group leader stood at the front of the room to guide the discussion through the use of specific questions that elicited class response. Two scribes wrote student comments on the board as appropriate. The board notes evolved as follows:

What do Betsy and Gunter want?

- Work less/earn more (less stress, concern for retirement)
- Financial stability and security
- Resolve uncertainty
- Get $1.5 million

How to get there?

- Older (60's), but no trained successor
- Successful business
- Developed big expectations of business value
- Developed unique patent
- Enthusiasm
- Frazzled

Is this a success?

Yes	No
• By feeling	Pie-in-the-sky asking price
• Gunter feels productive about business growth	
• Self-fulfillment	

Options for Betsy and Gunter

1. Fold business – exit
2. Make a business plan
3. New broker
4. Bring in partner
5. Status quo
6. Contact big companies that make tempeh
7. Look for companies that want to diversify
8. Hire employees to boost output
9. Look at marketing potential
10. Relocate to better market
11. Sell product direct, eliminate middleman
12. Hire consultant

Debriefing on "Betsy's Tempeh"

In case teaching, the leader has to be concerned with classroom facilities and environment. The blackboard ended up being very hard to read.

Teaching groups were satisfied with the discussions. Stang raised several key issues for follow-up to this situation:

- How would you value a small business?
- What is the legitimate role of a broker?
- What potential exists for commercial scale-up?

Peterson observed that the group made a classic case presentation, i.e., a whole group exercise, structured by leader's questions, filling the entire time. When a teacher senses that students are less involved in a case than they should be, he or she needs to question whether it is due to the students' preparation, the quality of the case, or the quality of the teaching. Any of these three factors can have an impact on how a discussion goes on a given day. Any participation approach gets old if you do not mix formats in the course.

Participants raised a question about the importance of bringing closure to the case discussion. Students certainly like closure, particularly when the teacher can share the decision made by the actual case participants. However, be careful that students do not confuse the actual decision with the "best" decision or the one right answer. Some open-endedness is good. It keeps the students thinking.

A question was also raised about the use of the real decision maker as a guest in class. This usually works best when the decision maker sits in the back of class during the case discussion and then critiques and debriefs the class afterward.

Mini-Lecture on Case Teaching Tips

To bring the whole teaching experience to a close, Peterson shared several tips about engaging in case teaching.

First, a "classic" structure does exist for teaching a case. It does not always need to be used, or used completely, but it helps to keep these steps in mind as a case discussion unfolds:

1. Define the problem or opportunity. (To do this, students must know who the participants are, what their goals are, and the gap between the goals and actual performance. Problems are undesirable outcomes, and the problem definition can often be the hardest part of any case.)

2. Determine cause of the problem.

3. Create alternatives.

4. Select the "best" alternative.

5. Develop a plan-of-action for implementation.

The theory and conceptual material taught in a course are most useful to the student in completing Steps 2 and 3 of the process.

Second, the case leader has to keep track of several critical things in the process of discussion:

1. The facts—what is known with certainty in the case

2. Assessments
   a. Those by actors in the case (may not be factual)
   b. Those by students

3. Linkages
   a. Between facts and assessments
   b. Between assessments

Effective decision making arises from effective discussion and understanding of these three elements. To the extent possible, bias needs to be driven from assessments while sound logic needs to build the linkages.

Finally, the case teacher faces numerous decision points as a case discussion unfolds. Only experience can help the teacher master these decisions and thus help students succeed in using the method. Peterson distributed Exhibit 1 from "The Case of the Lecture that Wasn't" (see part IB) which lays out the decision points critical to case teaching.

Systematic and Systemic Thinking Tools

By Ray D. William, Oregon State University; Robert Vertrees, Ohio State University

People tend to see the world systematically, systemically as relationships, or sometimes as a blend. Robert Vertrees describes a systematic diagramming tool known as concept mapping within the context of the "Karen" case. As members of Team C asked participants to identify key information from the case according to values, facts and problems or issues, Bob drew a first draft of the attached concept map and summarized it in the broader context of: (1) why he requires students to prepare a concept map or some other type of systematic diagram, (2) how concept mapping is based upon the "meaningful approach to learning," and (3) how a concept map is prepared.

Mind maps represent more relational or systemic thinking and diagramming. In class, Ray William has discovered that students begin to experiment with, and include, various types of systems diagrams in their class assignments. This begins when faculty describe a "lecture" as some sort of relational diagram, which is handed out as notes and explained. Students also discover that diagrams are a wonderful way to represent complex information in 10 minutes.

Choosing a tool or technique seems to depend on whether your expected outcome should be cause and effect or relational in nature. Steve Daniels at Oregon State University concludes that cause and effect diagrams, such as the fishbone technique, tend toward linear thinking, sometimes with a blame or fix-it mentality. In contrast, mind maps represent relational thinking patterns. We use mind maps as a central tool when describing complex and value-laden situations or issues supplemented with a linear outline of "notes." As we summarize, themes emerge and are prioritized based on the conversation. Discussion of each theme follows. As we enter the analytical phases, cause and effect diagrams emerge as central tools.

A Concept Map of the "Tom and Joan Karen" Decision Case

By Robert Vertrees

In a concept map, the main characteristic of any separately mapped concept is usually expressed by nouns or pronouns as single words or in phrases. I instruct students to place the concepts in rectangles, although some concept mappers use other types of figures. Relationships between or among concepts are written along the lines or arrows that link concepts. The main characteristic of any given relationship is usually expressed by verbs or some other nouns or pronouns. (Before proceeding, please scan the concept map on the next page.)

Why I Require Systematic Diagramming

The courses I teach are interdisciplinary, subject-matter courses. They are upper-division, dual-level, or graduate-level courses. Each course covers a wide variety of interrelated topics. For effective learning to take place in such courses, each student must be involved in the process of learning about these topics. This means that students must be provided with opportunities to gain experience in consciously and systematically developing cognitive skills in organizing, integrating, and depicting relationships among a complex of interrelated concepts about a topic.

A MODIFIED CONCEPT MAP OF THE "KAREN" CASE*
(BY ROBERT VERTREES)

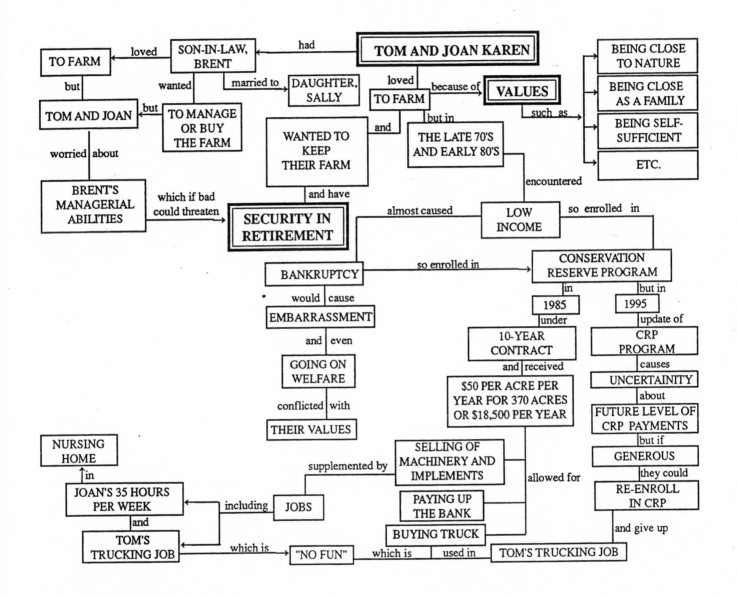

*The concepts in double boxes indicate the decision makers, their values, and their goal.

With this purpose in mind, I have prepared detailed written instructions for what I call a "Guided Learning Exercise (GLE) in Preparing a Concept Map or Other Systematic Diagram." Besides a concept map, students are given the option of preparing another type of systematic diagram, such as those described by Wilson and Morren (1990). However, most students choose to prepare a concept map. I require at least 12 interrelated concepts to be depicted in a concept map or other diagram. The oral or written topic that the map or diagram represents varies with the course offering. Among other sources, topics come from: (1) materials I specifically assign for this exercise, (2) a student's choice of any written or spoken material (e.g., from a reading, lecture or guest speaker), and (3) what a student says or writes in response to another course exercise.

Underpinnings from Educational Psychology

Teaching through the use of concept mapping is a meaningful learning approach that is based upon the cognitive learning theory of Ausubel (Ausubel, 1963; Ausubel, Novak, and Hanesian, 1978; Novak, 1983). In the epigraph to his book, Ausubel (1963) says: "The most important single factor influencing learning is what the learner already knows. Ascertain this and teach him accordingly." Of course, meaningful learning is very different from rote learning.

Preparing a Concept Map

In instructions to students, I distinguish between and provide examples of "true" and "modified" concept maps. The "flow of thought" or "flow of cognitive development" in a true concept map always proceeds down the page from the more general concepts to the more specific concepts. This "general-to-specific" rule applies the meaningful learning approach, because more specific concepts are given meaning by how they are related to the more general concepts that have already been placed on the page and in the mapper's cognitive structure.

If the "general-to-specific" rule is consistently followed, then concepts can be connected by lines instead of arrows because it is obvious that all the concepts and interconnecting phrases can be read downward. In a modified concept map, however, at least one or more (but not many!) of the interconnected pairs or sets of concepts are either: (1) at the same level of generality and, therefore, are placed directly across from each other on the page, or (2) involve a relationship that logically proceeds upward from a narrower concept to a more general concept located earlier on the map. In a modified concept map, therefore, some relationships between or among concepts cannot be read downward on the page. The words or phrases that represent the latter, broader types of relationships have to be placed alongside arrows (instead of lines) in order to indicate the proper direction of a relationship.

Grading the Concept Map

Of course, there is no right or wrong way to prepare a concept map. Each student represents any given set of complex, interrelated concepts in his or her own way. This is because the cognitive structure of one learner differs from the cognitive structure of each of the other learners (Ausubel, Novak, and Hanesian, 1978). This does not mean that there is no equitable and uniform way to grade the concept maps of individual students. I grade on the basis of: (1) the extent to which instructions were followed, (2) neatness, care in preparation, and evidence of proofreading, and (3) the extent to which the prepared concept map accurately represents the concepts and relationships of the oral or written topic that is being mapped.

References

Ausubel, David. 1963. *The Psychology of Meaningful Verbal Learning*. New York: Grune and Stratton.

Ausubel, David P., Joseph D. Novak, and Helen Hanesian. 1978. *Educational Psychology: A Cognitive View*. 2d ed. New York: Holt, Rinehart, and Winston.

Novak, Joseph D. 1983. "Concept-based Learning." In Kenneth E. Boulding and Lawrence Senesh, eds. *The Optimum Utilization of Knowledge: Making Knowledge Serve Human Betterment*. Boulder, CO: Westview Press, pp. 100-113.

Wilson, Kathleen, and George E. B. Morren, Jr., eds. 1990. *Systems Approaches for Improvement in Agriculture and Resource Management*. New York: Macmillan.

Mind Maps

By Ray D. William

Mind maps are a way of picturing what people are saying, sometimes feeling (Buzan). It shows relationships as people describe connections between parts of their story. Themes often emerge.

The technique encourages recall; it develops connections with the event to enhance memory. Some people say it's a wonderful way to study for exams and remember oral presentations.

To draw a mind map, listen for key noun(s) or verb(s) and relationships. Nouns go in circles; verbs along the lines. Pictures add life and feeling or a mood to certain aspects of the story. Adding lines to

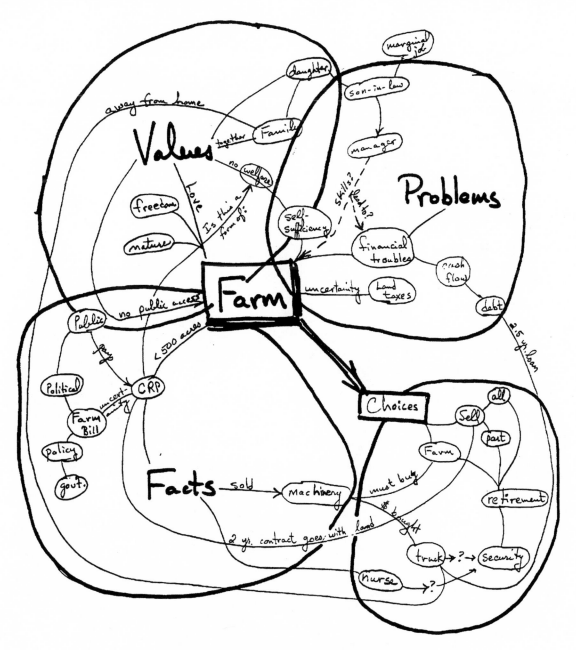

Mind Map of Karen Farm Dilemma

show relationships from one part of the story to another magnifies the utility of this technique. In group situations, we like to record group memory both in outline and mind map form. This approach enhances detail, records relationships, and encourages development of themes as described by participants.

During the conference, four themes emerged from the "Karen" case. This case was rich in values, facts, problems and choices with the long-term goal of retirement and security for the Karen family. Mind maps help focus discussion and keep learning on track while detail is recorded in an outline format. We use mind maps in classes and adult groups wishing to focus on various themes within complex, value-laden natural resource issues.

Reference

Buzan, T. 1993. *Use Both Sides of Your Brain*. Dutton, NY.

The Basics of Designing Cases into a Course

Group B Report

Facilitator: Chris Peterson, Michigan State University

The session began with an open-ended question: What issues does the teacher have to consider when designing cases into a course?

The group developed and discussed the following list:

- Course educational objectives
- Audiences for the course
- Sequencing of course materials
- Number and mix of cases
- Facilities and support
- Evaluation
- Course fit within total curriculum

Course Educational Objectives

Participants observed that the case method would be effective in achieving any or all of the following educational objectives:

- Applying knowledge and skill to real situations
- Discovering the need for special knowledge and skill
- Integrating knowledge and skill
- Grappling with complexity and ambiguity
- Developing teamwork skills
- Defining problems and opportunities
- Developing systems' thinking and action
- Dealing with pluralism and diversity
- Thinking critically
- Balancing ethics, economics, and science
- Discovering knowledge and skill
- Gaining experience in decision making

Audiences for the Course

Participants argued that cases would be effective with the following audiences:

- Professionals, professional societies, and business people
- Undergraduates at upper and lower levels as well as majors and non-majors
- Graduate students
- Industry/agency personnel
- Team teachers
- Colleagues
- Adult/Extension education

Sequencing of Course Materials

Based on material presented by Peterson, the discussion centered on five building blocks of case-course organization:

1. **Case-Concept Sequence.** When the teacher wants students to *discover the need for knowledge and skill,* open a course segment with a case and follow it with lectures and readings that develop relevant concepts and skills.

2. **Concept-Case Sequence.** When the teacher wants students to *apply knowledge and skill,* conclude a course segment with a case that requires the application of concepts taught in that part of the course.

3. **Case-Case Sequence.** When the teacher wants students to *discover knowledge and skills,* develop an all-case sequence in which students are continually challenged to discover the conceptual material that emerges from the multiple case experience.

4. **Case/Concept-Case/Concept-Comprehensive Case Sequence.** When the teacher wants students to *integrate knowledge and skill,* develop a series of course sequences that conclude with a comprehensive case solvable only by integrating concepts learned.

5. **Analysis-POA Sequence.** When the teacher wants students to *explore critical thinking and decision-making skills* in more depth, divide the case discussion into several days, focusing on analysis of the decision situation on one day, and making the decision with an associated plan of action (POA) on the second day.

The above building blocks can be mixed and matched to create a large variety of case-course designs depending upon the educational objectives of the teacher. For example, the "case sandwich" (case-concept-case) that combines building blocks 1 and 2 can be a powerful motivating and learning package for students.

Number and Mix of Cases

No magic number of cases exists. Anything from one case per course to cases in every session is possible. Cases might be used only to start a course (generate motivation for learning), and conclude a course (integrate course concepts). However, students gain more from the case method if they get to use and re-use it over time. It takes time and effort for both students and teachers to learn how to engage in the method effectively and efficiently. Therefore, a minimum of four or five cases is probably advisable. As with any teaching style, the teacher should search for cases with a variety of difficulty levels, subject content, and teaching settings. Certain subject areas also lend themselves more to the case method.

Facilities and Support

Case teaching is facilitated by classrooms with movable chairs, extensive blackboard space to track discussions, and good visibility between teacher and students and between students. A classroom arrangement that encourages an atmosphere of open discussion is preferred.

Teaching with cases also demands that a ready supply of cases exist or can be readily written. Copying and copyright permission, data access to supplement analysis, and networking of case writers are needed to assure access to a significant quantity of quality cases.

Evaluation

Unfortunately, time began to run out as this topic was discussed. The primary point discussed was that grading of participation is critical to the method's success. Several participants made it clear that participation grading really means contribution grading. Grading should be based on quality of contribution, not "air time."

Course Fit within Total Curriculum

Participants readily came to understand that relegating case teaching to one course in a curriculum is hardly optimal. Multiple course use is probably needed to develop the breadth of faculty and student skills needed to use the method effectively. This, in turn, raises issues concerning colleague reactions to case use, time commitment to write and teach cases, development of appropriate case publishing outlets to support promotion decisions, broadening the traditional notion of scholarship to include case research, and the interrelationship of courses within a curriculum.

Group C Report

Facilitator: Steve Simmons, University of Minnesota

Overview and procedure

This session began with the question, "What questions and decisions would you face if you were to design a course using decision cases?" After a brief period of small group discussion, responses were solicited. The remainder of the session was spent responding to the questions raised and soliciting ideas from the group.

Process

The courses suggested for using decision cases ranged widely from a freshman/sophomore introductory course in agricultural economics to a senior-level course in environmental restoration. Specific questions regarding how to design cases into courses also ranged greatly, but most centered on deciding how many, and what types of cases, to include. There was a strong interest in the question of workload and pacing of courses in conjunction with cases. We described several options for sequencing cases in courses in relation to other course teaching methods and activities, including:

- case—>lecture
- lecture—>case
- case—>lecture—>case (the "sandwich").

The group spent considerable time discussing the outcomes that cases can achieve in a course, including synthesis and application of knowledge, building awareness of ambiguity and complexity, creating a "need to know," and promoting critical thinking and student responsibility. Seldom is the specific outcome of the case critical to the course outcome, but the case serves merely as a vehicle for moving students to a higher level of understanding and capability for addressing higher-order outcomes. The facilitator emphasized that cases are merely another tool in our toolbox of education, albeit one that has been sorely neglected in the history of our respective disciplines in agriculture and natural resources. Cases are also a tool that can uniquely help achieve some outcomes that other methods (e.g., lectures) are poorly equipped to address.

Ozzie Abaye referred to a recent thesis completed under her direction showing the value of cases in helping students understand and apply concepts and information in a course.

Kim Mason, undergraduate student who had been recently instructed in a course using decision cases, commented on her perception of cases as a student and the inherent stresses with workload. She advocated that time be set aside in the course itself for students to work on and process the case prior to discussing it in any formal way. Simmons noted that the issue of "acclimation" of students to the process of case education had become important to him. He had found that students with little previous exposure to case teaching were sometimes confused about the instructor's level of expectation. Use of non-graded introductory (shorter) cases is often effective in taking the mystery out of the process in a way that is non-threatening and improves the subsequent performance of the students on later cases.

Assessment

Considerably more time could have been used in this session. There was time to hit some of the major points and consider the most obvious needs, but many of the more subtle issues were not addressed. The group did not even get to address all of the questions raised by participants at the beginning of the session.

Writing and Researching a Decision Case

By Emily Hoover, University of Minnesota;
Melvin J. Stanford, Mankato State University

Decision case writing is exhilarating. The method of writing is different than for professional, scientific journals. In a journal, issues, data, analysis and interpretation are all covered in the article. A case, however, contains only the issues and the data; the analyses and interpretation by the author(s) are put in a separate teaching note. Case writing should be conversational to hold the reader's interest and help him or her get into the case as a participant (in the role of the decision maker) rather than as a spectator. A few organizing tools can help both the experienced and the novice case writer with the case development process.

First, we will set forth an outline to help with initial organization of the case situation. This will be followed by a sequence of preparing a case and the related teaching note.

Outline of the Case

1. *The decision maker.* When thinking about developing a decision case, start with the decision maker. Who is he or she? What is the decision maker's organizational position and how did he or she get there? Give enough information so that the student can identify with, and vicariously, take the role of the decision maker.

2. *Background information* on the organization in which the case is set. What is the history of the organization? How does it fit into its industry? For example, if you were to write a case about an apple orchard, how does the orchard fit into the apple industry in the state and even within the country? Setting the stage for the students will help them to be brought into the thoughts of the decision maker.

3. *Issues, decisions to be made, dilemmas.* What are the issues facing the decision maker? How do these fit into the dilemma the decision maker

needs to resolve? Defining this part of the case is important in leading the reader into the decision-making process.

4. *Objectives of the decision maker.* What does the decision maker need or want to do? Where does he or she want the organization positioned after this decision is carried out? Delving into the objectives of the decision maker brings the case to life, gives it focus, and makes it easier for the student to vicariously become the decision maker.

5. *Alternatives or options of the decision maker.* What feasible options are there for the decision maker to achieve the objectives?

6. *Essential information for analysis, decision, support.* Include the basic information on which the decision maker did (or would) base a decision.

Outline of the Teaching Note

1. *Case objectives* (the beginning point for the case development process). What do you want to accomplish by using this case? Consider what cases are best used for: learning analysis, synthesis, decision making, and developing useful attitudes, judgment and wisdom.

2. *Uses of the case.* What course, seminar or program is the case designed for?

3. *Discussion questions.* These questions should be focused on the issues, primarily from the decision maker's viewpoint. Avoid clinical or "what do you think?" kinds of questions.

4. *Analysis by the author(s)* of the discussion questions and issues. This is where the author(s) takes a stand and interprets the issues and information in the case and sets forth the kind of work expected of the students.

5. *General conclusions by the author(s).* Consider what conclusions you would write in an article and what conclusions you would expect the decision maker (and well-prepared student) to reach.

Writing the Case

So you have gotten this far. Now that you have an outline, how do you write a case from it? Here is a suggested approach:

1. *Review your teaching objectives* for a class or research objectives for a project. Decide how your prospective case can help meet those objectives.

2. *Review any printed cases available on the subject.*

3. *Seek research support* (mainly time, but also costs). A good case requires as much work as a good paper or article.

4. *Reach agreement on access to data with the central decision maker or organization for the case (as in the outline).* Obtain a written release for unrestricted educational use of the case when it is complete. Most individuals and private organizations are sensitive about personal or financial data. However, a case can often represent certain decisions and issues without disclosing sensitive data. Where it is important to have sensitive data for the authenticity of a particular case, you may be able to disguise the data without distorting the realism of the issues and the decision situation. Published and public data usually don't need clearance.

5. *Interview the decision maker,* if possible. Try to obtain copies of written information pertinent to the issues revolving around the case.

6. *Select the information to be conveyed* to students or other case users. Quantitative data, and some kinds of background information, make good exhibits. Qualitative data make good material for the narrative body of the case.

7. *Arrange the exhibits* in the approximate order in which you want to refer to them in the case.

8. *Write a draft* of the case with reference to the exhibits.

9. *Write a draft of discussion questions* you want to assign the discussants and your interpretation of those questions, using only information you have included in the case. This will be the basis for the teaching note.

10. *Rewrite the case.* Then write the teaching note. As the case takes shape, gaps in the logic, data, or the flow of information will usually show up, and it will be necessary to get additional information to fill those gaps. At some point you will have to stop your additional research and say, "Enough!" That point will be reached when your case presents a realistic situation in which a decision can be made.

11. *Review the case with the decision maker* (usually not the teaching note), edit or revise it to his or her satisfaction and obtain written release for unrestricted educational use. Try to get a release for the case without disguise; however, disguising a case is better than abandoning it.

12. *Test the case* in a classroom as a regular assignment or in a discussion with colleagues. Then revise it based on those discussions.

Now you should be ready to submit the case and teaching note for peer review and publication. Some of the organizations that publish cases are: the *Journal of Natural Resources and Life Sciences Education, HortTechnology,* and the *Case Research Journal.*

Case research and writing doesn't always fit into this particular sequence. Some flexibility is needed, but the process has been proven time and again.

Appendix A2 provides 47 case outlines that participants at the conference prepared as part of the session on researching and writing cases. In addition to decision maker, background, issues, objectives, and alternatives, the original outlines also included essential information and exhibits planned. Some case outlines were excised from the proceedings at the request of their authors.

Reference

Stanford, M. J., 1992. "How to Prepare a Decision Case," in *Decision Cases for Agriculture,* eds. Stanford, M. J., R. K Crookston, D. W. Davis, and S. R. Simmons. College of Agriculture, Program for Decision Cases, University of Minnesota. (See especially papers 3, 4, 11, and 12.)

Managing the Case Experience

I. Student Assessment and Learning Styles

Facilitator: *Bill Welty, Pace University*

Summary by: Ray D. William and Anita Azarenko, Oregon State University;
Marion Brodhagen, University of Wisconsin

Overview

Two issues that span higher education include:

1. Assessment: how to evaluate learning and grade students doing cases.
2. Diversity: how to integrate diverse backgrounds, learning styles, abilities, and learning experiences in the classroom.

Process

Groups of three to five people discussed the development of a grading system for a course that had been assigned by the department head. Course objectives included teamwork, problem-solving and subject matter.

A student enrollment list suggested international diversity, as well as learning or interest diversity. How were we going to evaluate student learning and develop experiences that would facilitate learning within a mixed group?

Outcomes

Following 15 minutes in groups, assessment tools were listed and described:

Tool/Practice	Objective/Discussion
* Body count	Group projects require attendance.
* Participation	Develops preparation, oral, listening, and group skills.
	- Count verbal responses on scale of 0-3.
	- Assess quality of listening and talking as a contribution to the subject or discussion.
	- Choose quiet students and non-participants and alert them to be prepared to lead discussion at next session.
	- In the United States, participation often conveys an expectation of verbal response. What if we chose *contributions*? Would quiet students or foreign students contribute to group needs? Would individuals who process information orally discover other ways of processing? Which words are more inclusive or encourage diversity?
* Quiz/freewrite	Encourages preparation and assesses individual understanding prior to, and after, group discussion.
* Written papers	Enables students to summarize and determine their group of concepts.
* Small group activities: written/oral, final product	Develops group process skills and provides opportunity to discover and affirm knowledge
* Test	Evaluates knowledge of specific subject
* Team vs individual case inquiry	Discussion focused on cases and team inquiry, but should students be expected to participate in groups without instruction on group process? Perhaps cases are effective learning strategies for individual inquiry? Where should

students gain basic group facilitation skills?

* Tag team vs integrated teams

Briefly, discussants pondered tag team teaching and students' concerns about failing to get to know the professor. Are there other models, such as basketball-team-teaching that represent a well-integrated approach.

* Assessment and grading

Grading or assessing diverse contributions, learning styles and improvements presents a major dilemma for faculty and teams. Discussion focused around a blend of faculty and student team/peer evaluation. Although peer evaluation by students can become manipulative, and therefore questions of ethics and legality arise, two faculty reported ways to involve ownership by students. Peterson described a student-designed contract of policy and procedures for daily use and evaluation validated by faculty. Welty described a self-assessment of strengths and weaknesses followed by inventories.

Assessment

This session generated ample discussion. Discussion even followed the session. Welty suggested that this topic requires more scholarship and assessment of how effective we've been at self-evaluation. Cooperative learning literature is a good source of information and experience.

II. Enrichment Strategies

Facilitators: Steve Simmons and Kim Mason, University of Minnesota; Jack Stang, Oregon State University

Overview and Procedure

This session was conceived to be an "idea-sharing" time since several attendees had significant experience with case teaching. The session was divided into three components:

- enrichment strategies that add reality to cases,
- enrichment strategies for soliciting response to a case,
- concept of student-developed cases.

Process

Jack Stang began the session by describing his use of visits to Oregon farms in conjunction with decision cases. He noted that the use of cases prior to a visit often improved the students' understanding of the farm and sharpened their observational skills during the visit. One of the participants said that his institution had used field trips extensively in a "capstone" course, but had not yet used decision cases in conjunction with the field trips. He felt the idea had considerable merit.

Steve Simmons and Kim Mason shared an alternative "reality check" strategy—a conference call to debrief the case with the actual decision maker, after class discussion of the case. Mason, an undergraduate student who had taken a course at the University of Minnesota that used decision cases extensively, noted that the conference calls added considerably to the reality of the case and validated many of the points made by the instructors.

Simmons showed two videos that demonstrated the use of audio-visuals to enrich and add reality to the case experience. One of the videos was prepared specifically to accompany a case and the other was a commercially available video that had been edited for use in assigning the case.

There was considerable discussion about options for having students respond to a case. Again, a video was shown that illustrated an oral response to a case in the form of a role play. Since Mason was one of the players in the case, she described the process by which she and her group prepared their video responses and what they had learned from it.

Finally, Mason described an assignment to develop a decision case that she and a small group of students had done in a recent "capstone" course. She described the pitfalls of trying to develop a case that is close to one of the people, and the difficulty of achieving objectivity and detachment. There was also fruitful discussion of how to deal with uneven participation and contribution by members of small groups. We concluded that there

are a number of strategies for such situations, ranging from asking the students to evaluate each other's participation, to permitting the team to delete the names of non-participating students.

Assessment

This was another session where more time might have been used—but it was a very fruitful discussion nonetheless. The participants willingly contributed ideas and asked questions. The inclusion of a student as one of the facilitators was especially effective for several of the discussion topics.

More on Course Design

Facilitators: Chris Peterson, Michigan State University; Dave Davis, University of Minnesota

Davis emphasized that cases are an important way to make courses more interesting for students and to increase ties between extension and industry. He cautioned that the case writer must be careful not to damage trust with industry people. Peterson stressed the usefulness of cases at all levels of instruction. He uses cases in introductory level, upper-level undergraduate, and graduate courses.

The group was asked to volunteer any burning questions that they had about course design. The following questions were raised:

- How do you manage the method with large classes?
- Are there various levels of cases?
- How do you manage the length of time devoted to a case discussion?
- How do you use "science" cases?
- How do you evaluate students in meaningful and efficient ways?

The leaders attempted to focus on each of these questions.

Class Size

Physical arrangements within the classroom are key to managing case discussions with large groups. It is essential that students and teachers be able to see each other. Tiered lecture halls are acceptable facilities.

Getting participation is more difficult in large classes. Peterson uses a daily written quiz to count toward participation and to provide adequate incentive for students to prepare for class. Equity of participation is a major concern. This can be addressed by a combination of written quizzes and oral participation. In addition, seating charts can help manage participation. Use of small groups or discussion teams can facilitate participation and evaluation of the team contribution.

Various Levels of Cases

A case is a case as long as it involves a decision maker. Therefore, cases can be written at any level of difficulty that might face a real decision maker. Broader, more complex cases are more appropriate for upper level courses, and very specific issue cases can be used effectively in lower level courses. Even "armchair" cases (fictional cases based on limited facts supplemented with the author's experience or knowledge) can have a place. They are probably most appropriate for lower level students. It is important to note that armchair cases are not legitimate research documents, as are field cases.

Group members were very interested whether or not there should be an answer to a case. Real-life cases rarely have only one legitimate alternative for the decision maker. Case teachers should push students to understand that decision situations are complex and one right answer is rarely possible. Students also need to know that just because there is *no one* answer does not mean that *any* answer will work. Students must be challenged to find a limited range of feasible alternatives for the decision maker. The presence of one right answer (either implied by the case or forced upon students by the teacher) will stifle the creativity of students and severely limit the case discussion.

The issue of using cases as exams also arose. At lower levels, short cases (one or two pages) can be adequately used as exams. The questions should ask for essay response on such issues as: What is the problem in the case? What caused it? What are the alternatives? What is the best alternative? At upper levels, and with graduate students, case exams should be done as case write-ups and/or oral reports that require analysis and writing outside of the classroom.

Length of Discussion

Participants were concerned about how a case teacher manages the time given to a particular case within the course. If a teacher has a relatively short time to devote to a particular case, the teacher has several options: (1) select a short case with very focused issues, (2) if using a longer case, assign highly specific study questions that focus on a limited number of important issues, or (3) emphasize some subset of the decision-making process (e.g., analysis only) rather than push for a complete solution. Generally, students become more efficient at case analysis as they do more cases. Therefore, more effective discussion can be achieved in a shorter time later in a course.

Managing length of discussion can also take other forms. Limited lecturing can establish understanding of the case situation before students are assigned to focus on alternatives or plans of action. Case discussions can often be broken into two days with the first day focused on analysis of the situation and the second day focused on solutions and plans of action. Teams can be assigned certain pieces of analysis to be done outside of class in order to more efficiently get key issues on the table during class time.

The question arose as to how to get students to prepare for class. Graded participation and some method of warding off public humiliation seem to be rather effective.

Concluding Discussions

A brief discussion centered on "science" cases where the decision maker was a scientist faced with some critical decision about research. Davis suggested that such cases can be effective with graduate students, but are of limited value with undergraduates.

As time was growing short, grading did not get much discussion. Peterson and Davis both distributed sample syllabi to show course structure and grading policies actually used within several courses (see next eight pages). One final observation about course design: teachers must recognize that using cases does tend to reduce the content coverage of a course. However, the loss of content is usually more than made up for in terms of added student command of the content actually covered.

FSM 429   **AGRIBUSINESS MANAGEMENT**

Spring Semester 1995   Dr. H. C. Peterson
Time: M 12:40-2:30; WF 12:40-1:30   Office: 2A, Agriculture Hall
Room: 212 AGH   Phone: 355-1813
                          Graduate Assistant: Al

Wysocki

I.   <u>Required Text</u>

      <u>Strategy Formulation and Implementation, 6th Ed.,</u> Thompson and Strickland, Irwin.
      Case Packet at MSU Bookstore.

II.   <u>Course Objectives</u>

     1.   To develop students' knowledge and experience with business strategy.

     2.   To serve as a capstone experience for students' agribusiness management education.

     3.   To prepare students for the transition to the world of work, including emphasis on managerial decision making, managerial communication, leadership, and teamwork.

The focus of the course will be firms and industries in the agribusiness sector of the economy.

The attached course outline gives you a tentative daily list of the topics to be covered.

III.   <u>Student Responsibilities</u>

Lectures, readings, exercises, and cases will be used throughout the course. You are responsible for all in-class material presented and for all assignments of out-of-class work. If you miss a class, it is your responsibility to get lecture notes from colleagues and talk with the Professor or Graduate Assistant about assignments due.

Participation in class discussion is absolutely essential for successful completion of the course. It is only through participation that you show daily command of the issues raised by the course material. Participation in class is voluntary; a preference list will be used to encourage participation. Every student gets a participation grade for every class. This grade consists of points on the following scale: 1 = present but not participating, 2 = minor but useful contribution to discussion, 4 = substantive contribution to discussion, 7 = sustained and constructive contribution to discussion, and 10 = unusually insightful contribution to discussion. The first four absences from class receive a 0 participation point for each absence. After four absences, -7 participation points are assessed for each absence. The person with the highest number of participation points at the end of semester will receive a 100% participation grade; perfect attendance will receive a 72% grade; all other grades will be assigned using a linear function around these two points.

Academic dishonesty will not be tolerated. Failure of the course may result from such dishonesty.

There are no traditional exams in this course. Command of knowledge and course concepts will be shown through a number of written and oral reports. You are responsible for applying what you have learned to actual business problems. In addition to quality of analysis and synthesis, quality of presentation (written or oral) will be part of the grading. This course meets the upper level writing requirement.

The Team Project will involve a series of business research activities, written reports, and oral presentations. Projects will focus on different sectors of the grain and dry bean industry. Detailed assignments related to this project will be distributed at appropriate points throughout the course.

This course requires that students go beyond mere memorization of facts. Students must be willing to apply concepts and models. They must maintain an open, receptive, and inquisitive attitude toward learning.

IV.   Office Hours

W,F 1:30-3:00 p.m.
Other times by appointment

Students are encouraged to talk to the Professor or Graduate Assistant about course-related problems or concerns. Getting help early is essential. Feedback on the course's progress is appreciated.

V.   Grading

Participation	20%
Written Exercise 1	10%
Written Exercise 2	15%
Team Project	55%
TOTAL	100%

Each activity will be given a percentage grade. All written projects (except for the Round 3 team papers) can be rewritten for reevaluation of grade. The rewrite must be submitted within two weeks following the return of the initial grade.

The 4 point grades will correspond to the following percentages:

4.0	92-100%	Excellent Performance
3.5	87-91%	Very Good Performance
3.0	82-86%	Good Performance
2.5	77-81%	Adequate Performance
2.0	72-76%	Acceptable Performance
1.5	67-71%	Marginally Acceptable Performance
1.0	60-66%	Weak Performance
0.0	<60%	Unacceptable Performance

FSM 429 AGRIBUSINESS MANAGEMENT
COURSE OUTLINE

I. Business Strategy: The Basics

January	11	Course Introduction
	13	Overview of the Strategy Process (T&S Chapters 1 & 2)
	16 (2)	The Competitive Forces; SWOT Analysis (T&S Chapters 3 & 4)
	18	Mission, Objectives, and Strategic Posture (T&S Chapter 5)
	20	Selecting Strategy (T&S Chapter 6)
	23 (2)	Combs Nursery and Landscape--SWOT Analysis/Strategic Plan (Case) Team Formation

II. Business Strategy: Deeper Insights

	25	Assessing Finance and Operations
	27	Assessing Marketing and Personnel Ceiba-Geigy (Case)
	30 (2)	Scenic Hills Cooperative--SWOT/Strategic Plan (case)
February	1	Assessing the Competition; Seeing Strategy in Numbers
	3	Molecular Genetics
	6 (2)	Farmland Industries (Case); Written Exercise 1 Assigned
	8	Strategy as Spoken and Written Analysis
	10	Group Process and Leadership
	13 (2)	Written Exercise 1 due with discussion
	15	Implementing Strategy I (T&S Chapter 9)
	17	Implementing Strategy II (T&S Chapter 10)
	20 (2)	Case Exercise in Implementation; Team Project Check-In

III. Vertical Integration Strategies in Agribusiness

	22	Evolving Vertical Integration in Agri-Food
	24	Current Forces Supporting Vertical Integration
	27 (2)	Round 1 Team Presentations and Papers
March	1	CPC International (Case)
	3	Riceland Cooperative (Case)

----------Spring Break----------

March	13 (2)	Global Agribusiness Integration
	15	Charoen Pokphand (Case)
	17	The Future Agri-Food System; Written Exercise 2 Assigned

IV. The Reality of Strategy: The Grain/Bean Sectors

	20 (2)	Guest Speaker/Project Work Day
	22	Guest Speaker/Project Work Day
	24	Guest Speaker/Project Work Day
	27 (2)	Written Exercise 2 due with discussion
	29	Guest Speaker/Project Work Day
	31	Guest Speaker/Project Work Day
April	2 (2)	Guest Speaker/Project Work Day
	4	Guest Speaker/Project Work Day
	6	Guest Speaker/Project Work Day
	10 (2)	Round 2 Team Presentations and Reports
	12	Round 2 Team Presentations and Reports
	14	Synthesizing an Understanding of the Grain/Bean Sector
	17 (2)	Round 2 Feedback
	19	Project Work Day

V. Capstone

	21	Group Process Evaluation
	24 (2)	Round 3 Team Presentations and Reports
	26	Round 3 Team Presentations and Reports
	28	Strategic Management in Your First Job; Course Evaluation

Final Meeting: May 2, 12:45-2:45 p.m.

AnPl 5060
Integrated Management of Cropping Systems
Spring 1995

Course Description

Case study and discussion considering integrated production management of selected agronomic and horticultural cropping systems in Minnesota. Emphasis on problem analysis, principle application and decision making involving the integration of disciplines.

Objectives:

Upon completion of AnPl 5060, students will have:

1. Improved competence and confidence in problem identification and in using technically sound, analytical approaches to problem solving.

2. Improved their abilities to exercise judgement and assess options in crop, soil and pest management.

3. Improved their abilities to use team approaches to problem solving and decision making.

4. Improved their abilities to describe and defend problem analyses and management decisions both orally and in writing.

5. Greater understanding of principal management considerations involved in a wide array of agronomic and horticultural cropping systems in Minnesota.

Grading

Each case study will require either a written analysis or an oral presentation and, depending on the case, will be an individual or group effort. Case decision presentations in a written format will be evaluated by the faculty using the criteria shown on the attached evaluation form. Group oral case presentations will be pre-recorded on video prior to class (equipment for recording may be checked out from Teresa in 209 Borlaug). The length of a presentation should not exceed 10 minutes. Oral presentations will be viewed and evaluated (see evaluation form) by the course instructors. All students in each group must participate in preparation of an oral presentation.

Class participation will account for 10 percent of the total grade.

Decision Cases for AnPl 5060

Case 1.   "Bobby's Broken Barley" - Decisions must be made following the sudden collapse of a barley crop canopy during grain filling. (NE)

Case 2.   "Those Rascally Rabbits" - The Parlimentary Commission on the Environment must recommend a strategy for controlling rabbit populations in New Zealand. (NE)

Case 3.   "Perkin's Farm" - A southern Minnesota corn and soybean producer must decide about an equipment purchase with multiple implications. (IW)

Case 4.   "Carpenter Orchard" - Students weigh ethical, business and public opinion considerations in deciding whether to use Alar on apples. (IW)

Case 5.   "Agricultural Manager's Dilemma" - Problem analysis and decision making typical of an agricultural manager of a farm processing perishable products from contract acreage. (GV)

Case 6.   "Mueller Farm" - The option of producing lupins as a protein supplement for a Minnesota dairy is considered. (GW)

Case 7.   "Kalmes Farm" - Decisions are made regarding shifts to a lower chemical input agricultural system. (IW)

Case 8-9.   "Red River Valley Crop Consultant" (A) and (B) - A commercial crop consultant advises a Red River Valley small grain/sugar beet producer on production decisions faced prior to and during the 1989 growing season. (GW and GV)

Case 10.   "High Nitrate Showdown at Clear Lake" (A) and (B) - A small Minnesota community must decide how to respond to nitrate pollution of ground water. (IW)

Case 11.   "Linderman Farm: A Fresh Vegetable and Herb Farm" - A producer of fresh vegetables and herbs considers options for marketing and small business development. (GW)

Case 12.   "Red River Valley Crop Consultant II" - Red River Valley crop consultant must advise a grower about responding to a chlorotic and stunted wheat crop. (GV)

Case 13.   "Peterson Farms" (contingency case) - A management strategy is developed for a diversified crops farmer in the suburban-rural zone. (IW)

NE = not evaluated
IW = individual written
GW = group written
GV = group video

INTEGRATED MANAGEMENT OF CROPPING SYSTEMS - CASE STUDY EVALUATION
INDIVIDUAL OR GROUP WRITTEN PRESENTATION

Percent Evaluation	Decription and Comments
15	Professionalism
5	Audience defined and addressed
10	Style and effectiveness
25	Writing
10	Written correctness (grammar, spelling, and sentence structure)
15	Organization (format)
60	Content of Decision
20	Reference to and critical evaluation of exhibits
20	Technical correctness
20	Thoroughness

Total Percent _____

Additional Comments:

INTEGRATED MANAGEMENT OF CROPPING SYSTEMS - CASE STUDY EVALUATION
GROUP VIDEO PRESENTATIONS

Percent Evaluation	Decription and Comments
15	Professionalism
5	Audience defined and addressed
10	Effectiveness (visuals, etc.)
25	Speaking
15	Organization (length no more than 10 minutes)
10	Speaking manner (clarity, eye contact, voice)
60	Content of Decision
20	Reference to and critical evaluation of exhibits
20	Technical correctness
20	Thoroughness (Evidence of group input)

Total Percent _____

Additional Comments:

Case Use in Extension

Facilitators: **Ray D. William, Oregon State University; Marla Reicks, University of Minnesota; Roger Becker, University of Minnesota; Paul Gessaman, University of Nebraska[1]**

Critical thinking skills, problem-solving and capacity building were cited by Extension faculty as reasons for employing cases or facilitated learning and action techniques. Cases can de-personalize an issue or internal conflict as learners grapple with complexity and reality. Educators facilitate interactive learning by asking adults to identify key issues while focusing on either decision-making or problem-solving skills. Faculty report that participants' values often become integral to learning and action.

Facilitated action was described by Extension faculty in two contexts. First, case studies required facilitation toward decision making or problem solving by adult learners. The second involved facilitating discussions focused on contentious agricultural issues or complex management decisions influenced by weather, global economics, environmental consequences, or stakeholder views. Participants jump into the issue and work toward improvement while recognizing multiple perspectives.

Extension faculty discussed the "silver bullet" approach to learning—the "expert model" of teaching facts and information. However, we recognized that research and creation of factual information is becoming a symptom of, rather than an asset to, the learning process. Several faculty mentioned mini-lectures or facts being integrated into active learning sessions. Learning moments occur when individuals or groups identify and request needed information. We seemed to agree that a cultural shift in education was occurring; that we have a responsibility to engage learners in the learning and action process while integrating or identifying relevant facts and information.

What makes this learning approach work?

Topic selection and learning approach: Someone mentioned that case studies may be a solution looking for a problem. Most Extension faculty agreed that cases are a tool or technique to engage learners in a decision- making or problem-solving process. Topics of immediate and passionate concern to participants work best. Careful and purposeful choices of learning approaches encourage ownership of a learning activity and its outcomes.

Time: Case discussions and facilitated action involving complex issues require "in-class" time. Cases offer personal appeal. Short videos like the TV news segment of "The Sparrow" can be brief (i.e., 10 minutes) with a question "What should you do?" Perhaps home study materials should be developed for technical information and continuous learning. In contrast, people who become engaged in active learning sessions often report satisfaction and personal discoveries not achieved in other learning environments. However, cases are not for everyone (see "Learning styles" on next page).

Diverse audiences/reading skills: Faculty reported use of cases with Eastern European farmers. Perhaps cases could be translated, adapted, or written for Spanish-speaking or other cultural groups. Visual or diagrammatic learning and participatory techniques are being invented in developing countries (participatory rural appraisal). Another participatory training strategy involved engaging farm workers in a discussion about pesticide safety while they examined a short-sleeve shirt, long-sleeve shirt, and rain jacket. Participants shared their knowledge, experience, and concerns about each garment. Interactive learning occurred for 60 minutes with considerable enthusiasm.

[1]For their discussion contributions, we thank Dave Chaney, Joe Conlin, Colette DePhelps, Paul Dietman, Steve Ford, C. William Heald, Emily Hoover, Steve Laursen, Sally Noll, and Bill Wikke.

Learning styles: Compared to lectures, cases and active learning strategies are a mixed bag for learners who exhibit an array of learning styles. Bernice McCarthy has designed a learning system that begins with identifying the task/need, analyzing alternatives, practicing problem-solving, and applying the knowledge. Both group and individual activities are blended and balanced. In adult groups, we often see *doers* on the edge of their chair saying "let's do something even if it's wrong" while *planners* sit back wondering "how can you do anything until you plan?" Some people (approximately 25 percent in North America) assimilate information best from lectures, while others prefer participation, problem-solving, seeing, etc.

Distance education: Questions of study via computer, Internet, home, and distance education contributed to several cycles of discussion and imagination. One person reported using case studies at three distance education sites with groups of 30, 40, and 15. Participants began to address each other by first name across sites. It sounded very innovative and successful.

Evaluation: One person reported using a numerical score of 1 to 10 for several key questions combined with open-ended questions. Another person developed an ordinal scale and series of questions to assess whether participants expected to act on workshop information. A follow-up assessment three to six months later asked whether they had used the information. Organizers assume that people who fail to return the questionnaire probably failed to apply the information. A third colleague reported asking "What was most useful, least useful, and what follow-up was needed?"

Scholarship: Comments throughout the discussion suggested that great opportunities exist in developing brief cases (i.e., the 10-minute Sparrow video), realistic cases for adults or farm families, facilitated action of issues, and practical learning or group facilitation techniques.

Assessment: Although time was brief (1.5 hours for Extension faculty!), a "round robin" closure suggested that expectations had been achieved. Probably more important is networking or knowing whom to contact for future information.

Integrating Decision Case Discussion with Cooperative Learning Strategies

By Emily E. Hoover, University of Minnesota

What is cooperative learning and how can teachers use it to increase involvement and learning in classes? More specifically, how can teachers use cooperative learning in the context of decision case teaching? These were the primary goals of this session.

Cooperative learning can help ensure that students actively create their knowledge rather than passively listen to yours (Johnson et al., 1991). Structuring learning situations so that students interact with each other is a prerequisite for creating a cohesive class striving to achieve mutual goals. Research has demonstrated that cooperation among students produces greater achievement and higher-level reasoning, more positive relationships among students, greater acceptance of differences, and higher self-esteem. The advantages of using cooperative learning techniques, especially as applied to case teaching, are immense. Faculty using the case method want students to work together, and not in isolation, to solve problems. However, most faculty have been taught and encouraged to prevent students from helping, talking to, or encouraging each other. Therefore, a shift in emphasis has to be made to design classroom experiences that enhance cooperation, and thus enhance learning for all students.

Basics of Cooperative Learning

What are the basic principles of cooperative learning? Putting students into groups is not the same thing as cooperative learning! According to Johnson et al., five elements must be included in a lesson plan in order for cooperative learning to take place: positive interdependence, face-to-face promotive interaction, individual accountability, social skills, and group process. Keeping these principles in mind, let's walk through the "Thea Grossman" case. The case is included in Chapter 7 of Part IB, and I encourage you to read it before proceeding.

Thea's dilemma centers around wanting more active participation in her class. She attempts to increase student participation, but the exercises flop. She is left wondering what she should do.

I began this session by asking participants to focus on social skills. The group discussed skills needed by a member of a small group. The list generated included: listen, be prepared, contribute, participate, take risks, respect each other, think, be flexible, negotiate, synthesize, be able to be comfortable with ambiguity, take and give control, ask questions, record information, check for understanding among group members, keep on task, keep on time. This list was generated in about five minutes and help set the tone for all participants to be involved in the rest of the session. Having participants generate the list gave them ownership. This allowed participants to remind each other when those skills were needed. A good example from one of the groups included an exchange that was getting rather heated. One member looked at the other and said, "You've not been using your listening skills. I need you to listen." The answer, "You're right, I should stop doing all the talking."

The group was then subdivided following a few guidelines to help ensure balance: no self-selection, a group size of three or four; when groups of three, try to have either two women and one man, or all women. If groups self-select, members are apt to bring in outside, personal agendas and not stay on task. Group sizes of four or more make participation of each group member more difficult. Making sure that women are in the majority in a group increases their participation within the group, particularly for undergraduate women. All the groups moved their chairs so that they were able to have face-to-face interaction and view the plan of action they were proposing.

Enhancing Participation, Individual Accountability

To structure positive interdependence, the group assignment was to generate a plan for how Thea could redesign the class to enhance participation. This also forced groups to think about their strategies and decide on what aspects were successful and which needed changing. Groups needed to agree on the plan and the strategies for solving Thea's dilemma. Resource interdependence was created by giving each group only one large sheet of paper and one marker. These sheets were then posted around the room for all groups to read.

Individual accountability is often missing in group work. To make sure that all continued to participate, participants individually viewed other group's answers and decided which were "best." This gave participants two added benefits: first, they had been sitting for almost an hour and this activity made them get up and move; second, they had a chance to see how other groups answered the dilemma. Each group took a different tack and no one answer had included everything.

Back in small groups, positive interdependence was structured into the discussion. Participants had to come to a consensus regarding the best answer. Each group reported the answer it had chosen and explained why. This technique increases the interest for all the others in the room since it focuses the discussion on *why* the answer was chosen, not on re-explaining the answer.

Processing at the end of the exercise was done by answering the following questions: What three things did you do well as a group? What one thing would you do differently next time to help facilitate group learning? It is imperative that groups do this exercise each time they meet. It brings closure to the exercise, and for participants, it focuses on the positive work of the group and how each person could change his/her behavior to help in group learning.

The cooperative learning lesson plan that I prepared for the session follows. To make cooperative learning meaningful to the participants, advance planning is important. The attached sheet makes sure that I, as a facilitator, think about the facets of the exercise that ensure cooperative learning. This allows for the session to run seamlessly from activity to activity.

Literature Cited

Johnson, David W., Roger T. Johnson, and Karl A. Smith. 1991. *Active Learning: Cooperation in the College Classroom.* Edina, MN, Interaction Book Company. [7208 Cornelia Drive, Edina MN 55435 (phone: 612-831-9500)].

Lyon, Sally, 1992. "Thea Grossman." In R. Silverman and W. Welty, eds, *Case Studies for Faculty Development*, White Plains NY: Pace University [78 N. Broadway, White Plains, NY 10603 (914-422-4321)].

Cooperative Lesson Planning Form

Grade Level: *faculty at seminar*

Subject Area: *active learning intro*

Lesson: *introducing some principles of active learning and how to use those principles in classes when working with decision cases.*

Objectives

1. Academic: *how to prevent active learning strategies from falling apart in a class;*

   understand the benefits of small group and individual participation within the context of a large group.

2. Social: *to have participants engaged and contributing to the group project.*

Decisions

1. Group size: *3, with maybe one group being 4*

2. Method of assigning students: *count off (with 36, need to count to 12; with 40, 13)*

3. Roles: *task master, time keeper, encourager/checker*

4. Room arrangement: *move chairs so that every one in each group is in direct eye contact.*

5. Materials:
   __X__ a. one copy of the questions being asked per group
   _____ b. jigsaw
   _____ c. tournament
   __X__ d. one copy of the case per person
   _____ e. other

Explaining Task and Goal Structure

1. Task: *To redesign the assignment that Thea gave to her class to make sure that the class is productive and not a total disaster.*

2. Criteria for success: *If everyone comes away from the sessions with an idea to implement in his/her classes, whether or not it deals with decision cases.*

3. Positive interdependence: *Will be structured by chair arrangement and by providing one set of questions for each group.*

4. Individual accountability: *Individuals will be called on during the large group discussion. They will be told ahead of time that this will happen. May collect writing.*

5. Expected behaviors: *Everyone should be in the discussion when in small groups.*

6. Intergroup cooperation: *For this exercise, not much cooperation between groups will be expected other than sharing discussion afterwards.*

Monitoring and Intervening

1. Observation procedure: _____Formal
   ___X____Informal

2. Observation by:___X___Teacher
   _____Students_____Visitors

3. Intervening for task assistance: *Will not intervene, will listen-in on conversations among groups to make sure that they are on task and everyone is contributing.*

4. Intervening for teamwork assistance: *Not likely any will be needed.*

Evaluating and Processing

1. Assessment of members' individual learning: *Through large group processing (see below).*

2. Assessment of group productivity: *If each group finishes the task.*

3. Small group processing: *What three things did your group do especially well? What one thing could you do to help facilitate group work?*

4. Large group processing: *What one thing did you bring away from today's discussion that you can use in courses you teach?*

5. Charts and graphs used: *Instructions both orally and written; comments from the groups written on sheets of paper which will be posted for all to see.*

6. Positive feedback to each student: _____
   _____
   _____.

7. Goal setting for improvement: _____
   _____
   _____.

8. Celebration: *MAKE SURE TO CONGRATULATE THE GROUP AS A WHOLE FOR ALL CONTRIBUTIONS MADE.*

9. Other: _____
   _____
   _____.

Adapted from: Johnson, David W., Roger T. Johnson, and Karl A. Smith, 1991. *Active Learning: Cooperation in the College Classroom.* Edina, MN: Interaction Book Company.

Case Teaching Issues for Experienced Teachers

By Josef Broder, University of Georgia; Kent Crookston, University of Minnesota

Case teaching is full of surprises, some pleasant, some we would like to avoid. By design, case teaching is a dynamic process where outcomes depend largely on the quality of interaction among participants. As facilitators of the case method, teachers should influence or guide case discussions in a measured and unobtrusive fashion. Lacking the control mechanisms of traditional lecture methods, case teachers face a different set of issues than their traditional counterparts. This session examines selected issues faced by case teachers. While the perspective is from experienced teachers, the issues are particularly relevant to novices of case teaching. The following issues grew out of the personal experiences of the authors and are not intended to be all-inclusive. These topics are not presented in order of importance, but according to the flow of discussion during the session.

Know Your Audience

An old adage in public speaking is "know your audience." While important in traditional teaching styles, this adage becomes even more critical in case teaching. Case teachers need to know individual students and not just generalities about the audience. Knowledge about individual students should be sufficient so that the case teacher can interact comfortably and meaningfully with students. In lecture, the speaker can ask rhetorical questions and not really care about the response or respondent. By contrast, the case teacher cannot tune-out the dialogue, but must build on individual responses and encourage participation from others.

How does the case teacher learn enough about his/her students to be an effective facilitator? First and foremost, the use of name tags serves to break-the-ice among facilitators and students. Unless students know one another, they may have more diffi-culty interacting with other students than with the instructor. Name tags encourage interaction by adding a name to a face, and oftentimes, a face to a comment. Some will view the use of name tags as childish and prefer to stay anonymous, at least at the outset. Your role as facilitator is to reassure the students to be active participants and not just nameless spectators.

Second, assigned seating and seating charts can help the facilitator remember students and comments. To facilitate this process, position name tags before the participants arrive, or use seating charts. Of course, some students will suspect the session is being overly controlled or staged. A seating chart can also help the facilitator connect people and comments.

Third, students can be asked to stand and give their names when they speak. This technique is useful in classrooms with poor lines of sight and acoustics, or for individual introductions at the beginning of the session, assuming the class is not too large. The practice can be used throughout the session. In large groups, students may have trouble remembering names and enforcing this practice becomes difficult as the students interact more frequently. An element of spontaneity is also lost in the process.

Fourth, the case teacher can request and review biographical sketches of the students before the session. This is especially important for first- or one-time case sessions. While student bio sketches may not predict which students will readily participate in the discussion, they give the facilitator some familiarity for asking particular questions. Ideally, the facilitator may want to observe the students in class or in a previous case to get a feel for the responsiveness of various individuals. While such prior information seems contrived, it does allow the facilitator to be more effective in directing the discussion. Knowing the type of response that selected individuals are likely to offer can help the facilitator identify students for opening, transition, and closing comments.

Difficult Students

The greatest challenge facing the case teacher is transferring the responsibility of learning from the teacher to the student. As teachers and students move out of their traditional comfort zones, a certain amount of anxiety is created. Teachers must develop a sense of trust in students and in the process of teaching by the case method. Students have to actively participate in case discussions in a way that is new to many of them. Some students make this transition rather easily, others experience difficulties. These difficulties arise when students talk too much, or too little, are out-of-sync with the discussion, or are antagonistic toward others. For a case teacher, these difficulties fall under the topic of difficult students.

Difficult students often stem from differences in expectations. Teachers have expectations about how they want the discussion to develop, what students should do, what makes a good case participant. To ensure that the discussion meets the teacher's expectations, a set of ground rules should be established and communicated at the outset. For example, should students be required to raise their hands before speaking and stand up while speaking? Should students be allowed to interrupt other speakers? Should students address the teacher or other students? How long and often should an individual student be permitted to speak? These ground rules should be spelled out and then followed during the case presentation. Of course, these ground rules are not an end in themselves, but are intended to promote and encourage class participation.

Difficult Situations

Often overlooked or beyond the teacher's control are the physical facilities. Large rooms with fixed desks, filled to capacity, can be especially challenging, if not disastrous, for the case method. An auditorium-style room might be appropriate for an instructor-centered case, but would be wrong for a student-centered case. For good interaction, students must be able to see, talk to, and hear each other. Without the proper room setting, full participation may not be feasible or desirable. If you know this ahead of time, consider starting the case with small group sessions or discussing the case in a larger setting using fishbowl techniques, whereby you lead the case to a select group while others observe without speaking.

Classes can also be too small for effective case teaching. While students tend to get lost in large classes, small classes often lack the synergism and stamina of larger groups. In small classes, students have to speak more often and are more noticeable when they are silent. When using cases in small groups, recognize that there are limits on the number and duration of cases that can be presented.

Role of Gender

The case method attempts to create an atmosphere of willing participation. This assumes that learning is a shared responsibility and a democratic process. While participants are on an equal footing, they are not completely divorced of roles they play in society. Gender, age, and other personal characteristics can affect the tone and frequency of interaction. Case teachers should be sensitive to personality differences. As you present cases in different settings, you may notice, for example, that men participate more aggressively than women; that younger students are reluctant or embarrassed to talk; and some nationalities will never speak unless asked a direct question. While teachers should be cautious about forming stereotypes, they should be aware that some groups are more willing to participate than others, and that some individuals or groups need a little coaching/encouragement to participate.

Case teachers should also recognize that their own personal traits can affect the discussion. A number of questions were raised in the session. How might the character of the discussion differ between a male and female facilitator? How do gender differences between the facilitator and participant affect willing participation? Do age differences matter? How might the quality of discussion be affected when the facilitator is in a subordinate or superior position to the participants?

Handling Criticism

Interesting cases involve characters with differing opinions and personalities who confront problems that don't have obvious answers. The student participant often identifies with certain characters, and is critical of others. Since participants offer different perspectives, criticism of characters or their actions can be taken personally by other participants who share these views. While differing opinions make for lively discussion, be on guard for opinions that stifle rather than encourage interaction.

When differences of opinion become unproductive, the facilitator can try a number of techniques. First, critical comments stem from participant frustration, which can sometimes be alleviated by acknowledging the comment or by probing the participant. If this fails, the comment could be redirected to other students. Second, the discussion can be redirected from the perspective of a different character in the case, perhaps one who gives perspective to the criticism. Third, the discussion can be shifted away from opinions about who is at fault toward finding facts and details of the case setting. An extended period of fact finding often allows participants to develop a rapport without becoming prematurely antagonistic.

During some case presentations, students may be critical (skeptical) of the case or the facilitator. Students are apt to criticize cases they view as being contrived or unrealistic. At the outset of the case, students may discount the relevance of the case and become unwilling participants. If this happens, the facilitator should be patient and give them time to become familar with the case or the case method. Here, the facilitator might encourage the group to identify sources of the problem presented in the case. As discussion begins, ownership or acceptance of the case should follow and a state of suspended disbelief will emerge.

Criticism of the facilitator results when students view the teacher as part of the case problem. To address this, the teacher should redirect the case to the participants and play a more passive role in the discussion. This might involve asking more questions and avoiding declarative statements or opinions about the case. Some of this potential criticism can be addressed in ground rules by stressing that there are no right or wrong answers, and that the facilitator doesn't have any more details about the case than the students.

Dealing with Silence

Silence plays an important role in case teaching. Silence is often more excruciating for faciliators

than for participants. In case teaching, there are periods of collective silence; the class has nothing to say and you are at a loss as to how to proceed. To facilitators, seconds seem like minutes and the urge to talk or lecture intensifies. At times, the facilitator gets tired or distracted and has trouble redirecting the discussion. When silence prevails, the facilitator should not panic. Periods of silence are sometimes needed before the more reluctant students will speak. It is also an opportunity to direct the discussion toward students who have not participated. Students share a responsibility of discussing the case. It is not the teacher's role to do a monologue. If silence persists and panic ensues, redirect the questioning or change perspectives of the case. This gives students with different perspectives an opportunity to participate.

The facilitator has expectations about how individuals should participate. At one extreme, there are students who would rather not say anything at all. If silence by individual students is unacceptable, the facilitator should spell this out as part of the ground rules. Given the diversity of personalities in the classroom, some flexibility is advised. At issue is what constitutes meaningful participation in the case discussion. Some students learn by talking, others learn by listening. Still others don't learn when they are talking, and silence is a sign that students have tuned-out the discussion. If the facilitator expects everyone to participate orally in the discussion, he/she might try using "warm calls" as a way to encourage reticent students to speak. With warm calls, the teacher announces privately or publicly that he/she would like to hear from a particular student on a particular topic, sometime during the discussion. This creates an expectation for the student to speak, but leaves the student some flexibility as to when to speak. The warm call can be contrasted to the "cold call" whereby the teacher calls on a student without prior notice and with little time for reflection.

At the other extreme are students who talk (or want to talk) too much. Be suspicious of the student who keeps his/her arm raised for long periods of time. Some case teachers argue that when the arm is raised, the mind stops thinking. By the time you call on him/her, he or she may be responding to a comment made five minutes earlier. These out-of-sync comments can frustrate the discussion and the facilitator.

Body Language

The location, actions, and mannerisms of the case facilitator are key ingredients in case teaching. The student's willingness to participate in case discussions is influenced by the facilitator's body language. Students have comfort zones when interacting with others. Anecdotal evidence suggests that the facilitator can encourage comments by moving away from and keeping a safe distance from a student or discourage comments by moving overly close. Breaking eye-contact or turning your back on a student may also discourage comments, but such tactics should be used judicially. To encourage group discussion, the teacher can distance himself or herself from the class, be seated, or obstruct the line of sight between teacher and students. Body language can be use to probe individual students for more information; while questions directed to the facilitator can be redirected to fellow students.

Students are easily frustrated when their comments are not acknowledged by the facilitator or fellow students. Body language can be used to acknowledge the validity of individual comments. Techniques for acknowledging comments include listening, maintaining eye contact, probing, repeating or rephrasing comments, or writing comments on the chalkboard or flip chart. When rephrasing comments, be careful not to change the content, (i.e., put words in the student's mouth). Avoid the temptation to answer probing questions or give information not contained in the written case materials.

Managing different forms of acknowledgements can be challenging. For example, when writing a student's comment on a flip chart, the teacher loses eye contact with the student and the class. For some, writing and probing at the same time is difficult. Student comments must be condensed to fit into categories or space allowed on flip charts or chalkboards; thus the original meaning can be lost in the translation. Using multiple facilitators or a combination of facilitator and recorder allows at least one facilitator to maintain eye contact with the class while the other records comments. Some may find these joint efforts to be disjointed and unmanageable. Much depends on the level and sequencing of materials discussed in the case.

Professional Development

Case teaching is a far more dynamic activity than lecturing. The lecturer gets a sense of how students perceive him/her by rehearsing in front of a mirror. Case teachers are apt to learn little from this technique. To get a complete picture of how you facilitate a case, we recommend videotaping and peer observation. Unobtrusive videotaping can help the facilitator see the quantity and quality of interaction. Peer review of video tapes and direct observation by peers are useful for interpreting and diagnosing facilitator effectiveness. Last, but not least, serious students of case teaching should attend case workshops, such as this one, to observe and participate in case teaching sessions.

DOTS: A Group-Owned Visual Evaluation Technique

By Ray D. William, Fred Smith, and Larry Lev,
Oregon State University

Participants often complete pre- and post-conference questionnaires designed to inventory skills on registration forms or evaluate the conference or workshop in exit surveys.

Results are usually shared in a report or proceedings several weeks or months after the conference. To encourage participatory or group-owned learning, we decided to explore methods for integrating pre- and post-workshop inventories and evaluations as an interactive, visual part of the meeting. This complements other experiments that have explored participatory posters and systemic diagramming as ways to achieve clarity, brevity and ownership, while maintaining spontaneity and FUN. Nominal group technique using dots to prioritize topics or themes has worked well among

adults. Thus, adaptation and a bit of invention resulted in DOTS ("Delta over time system") as a technique for visually representing where individuals perceive themselves at the beginning and end of an event.

DOTS allows participants and organizers to gain immediate and visual access to:

1. pre-conference knowledge, skills, or attitudes of individual participants.
2. conference topics and themes.
3. post-conference assessment and other forms of closure.

At registration, most workshop participants rated their experience and understanding of decision cases as "little" to "moderate" (Table 1). To make this information visible to everyone, we modified the question to include understanding only and hung the poster near the entrance to the first plenary session. With their name badges, each participant was given two adhesive dots, one red and one yellow. Each dot carried the registrant's unique

Table 1: Understanding of decision cases before and after the workshop: Three measures
— DOTS poster, registration forms, and survey responses.

| | | Low ⟵ ⟶ High | | | | | |
		(1)	(2)	(3)	(4)	(5)	Total
DOTS	pre	18	32	21	8	2	81
	post	0	0	36	30	2	68
Registration	pre	21	9	20	4	6	60
Survey	pre	32	—	26	—	4	62
	post	0	—	30	—	33	63

identification number. Chris Peterson encouraged people to apply their red dot to the poster as they entered. Table 1 shows how participants rated their level of decision case familiarity with the dots, the registration form, and pre- and post-workshop evaluation survey forms. Participants were instructed to place the yellow dot with the same ID number to indicate their level of case understanding at the end of the workshop.

Participants could see how they ranked themselves relative to other participants (Figure 1). Most participants placed their dot between "little" and "moderate" understanding of cases and case-teaching techniques. Both the registration data and DOTS placement resulted in similar responses.

Post-conference results from both DOTS and exit responses show substantial movement toward greater "understanding of cases and case-teaching techniques" (Figure 1 and Table 1). A quantitative analysis was completed by replacing words with numbers (1-5) and subtracting the pre-workshop score from the post-workshop score. Mean improvement was 1.4 units toward the right. Only one person reported zero learning (meaning pre- and post-workshop DOTS remained in the same position).

Another post-workshop question was asked: "How would you rate this conference in terms of meeting your expectations?" Although descriptors varied somewhat, results were similar between the two assessment techniques. All DOTS were clumped on the right side (Table 2).

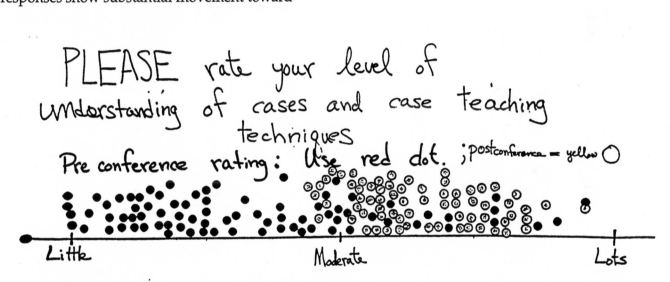

Figure 1: Participant self-ratings of "understanding of cases and case teaching techniques" using DOTS before (solid dots) and after (hollow dots) the workshop.

Table 2: How Would You Rate This Conference in Terms of Meeting Your Expectations?

		0		0		+60	
DOTS:	I———————		I———————		I———————		I
	Failed		Neutral		Met		Exceeded
Survey:	0		1		19		40
	Poor		Fair		Good		Excellent

Many people commented on the potential for peer pressure when people place their DOTS as part of a participatory group process. Evidence from anonymous survey questionnaires suggests that people respond honestly, according to how they feel. The overwhelming advantage of DOTS is making goals or evaluation criteria explicit, visual, and immediately available to all participants. In addition, we've found that DOTS forces organizers to distill and focus their efforts or objectives. Usually, four questions, or fewer, works best. By reviewing responses and the series of questions with the group, discussion and learning are both focussed and participatory.

Overall, DOTS is a quick, easy technique for meetings, field days or the classroom that improves communication, participation, and assessment. As with most inquiries, the greatest amount of time and effort is spent on designing relevant and clear questions. It works when you get the questions right.

Creating a Vision for Land Grant Universities in Their Teaching and Learning Programs

By Thomas D. Warner, W.K. Kellogg Foundation and Kansas State University

The goals of this presentation are (a) to identify trends that will affect higher education and our teaching and learning environment in the future, and (b) to present information on the W.K. Kellogg Foundation's funding initiatives to facilitate positive change within our land-grant institutions.

Higher education is coming under assault on many fronts. A large and growing number of individuals, including legislators, feel that universities are self serving and unwilling to change to meet the dynamics of changes in technologies, changes in our student populations with regards to ethnicity, gender and age groups, changes in the competitiveness within the higher education field, and unwilling to review the institution of tenure. This view has had direct negative impacts across our country's campuses.

Within the next 10 to 15 years, the changes that will take place within our higher education system will be profound. While residential campuses will continue to exist, the institution's branch campuses will be the student's home. Higher education institutions will provide courses on demand, not courses set up for the convenience of the instructing faculty. In the virtual university of the future, instructors will teach courses that will be seen by students across states, regions, the country and the world, and students will take a greater role in the learning process. The instructor will be a source of knowledge and a facilitator of learning.

A current trend is the shift back to excellence in undergraduate instruction. Major public, research-oriented universities are shifting toward student-oriented research institutions. Universities are also looking at more broadly defined scholarship to recognize excellence in the classroom.

As the virtual university moves toward our collective realities, the W.K. Kellogg Foundation has embarked upon a major funding initiative to support institutional change within the land-grant university system. Twelve project teams have been identified to participate in the food systems professional education initiative. These 12 project teams include 46 institutions within 22 states. The lead institutions are land-grant, however, a broad array of community colleges and some private institutions of higher education are also incorporated within this initiative. Following is the central focus and operational context of the initiative.

Issues Explored by the Food Systems Professions Education Project (FSPE)

With a challenge to envision a successful, long-term educational future, participants in this initiative are exploring a wide range of issues, including:

- Responsiveness to the varied goals, needs, and preferences of diverse audiences.
- Curriculum reform guided by consumer needs.
- Use of new or nontraditional educational technologies.
- Lifelong, continuous learning.
- University faculty rewards and incentives.
- Multiple sources of knowledge for the solution of problems.
- Collaboration and integration among disciplines for the practical application of knowledge.
- Collaboration and resource sharing among institutions within, and across, state boundaries.
- Long-term sustainability of food systems and higher education systems.

Who is Involved?

From higher education:

- Land-grant universities, other state colleges and universities, and community colleges.
- Educators, administrators, and students.

- Professional societies.
- Extension, research and teaching committees on organization and policy.

From the food system:

- Producers, retailers, and marketers.
- Business and industry representatives.
- Environmental and natural resource conservation organizations.

From the general public

- Diverse citizen and consumer interests.
- Citizen-based civic organizations.
- Youth leadership organizations.

Others:

- State and federal agencies and elected officials.
- Policymakers from local to federal levels.
- Philanthropic foundations and organizations.

FSPE Timetable and Anticipated Outcomes

The initiative is proceeding in two phases:

Phase 1: Visioning (Spring 1994-Fall 1995)

Common features of the visioning phase:

- Collaboration among two or more higher education institutions.
- A process of reflection, dialogue, and vision building.
- Expanded partnerships with business, industry, and communities.

Within these guidelines, each project has its own structure and process to build a vision unique to its own geographic and constituency needs. During the visioning phase, the partnering organizations in the FSPE initiative hope to achieve:

- A vision for food systems education, with implications for changes in land-grant universities and higher education in general.
- New structures for engaging citizens in vision building, decision making, and agenda setting.
- New models for educational responsiveness.

Phase II: Implementation (1995-2000)

The 12 Phase I projects are expected to develop implementation proposals based on their desired future or vision and submit them to the Foundation for possible Phase II support. Phase II consists of:

- Coordinated, collaborative action directed toward achieving the long-term visions.
- Sharing of resources among the participating institutions.
- Building the capacity to effect long-term change.
- Building partnerships with business, industry and community.

The implementation phase anticipates:

- Revitalized approaches to the education of food systems professionals.
- Systems changes guided by constituent needs and involvement.
- Professionals prepared to respond to the dynamic issues of 21st century global food systems.
- Networks of collaborating institutions.
- A critical mass of institutions, sufficient to bring about policy changes.
- An adequate, high-quality food supply.
- An improved quality of life for this generation and those to follow.

Maintaining the Momentum

By Eunice Foster, Michigan State University

The 2½-day conference ended on an extremely high note of enthusiasm, with a commitment to further advance the development and use of decision cases in the agricultural and natural resource sciences. The closing session had participants assess the effectiveness of the conference and share their ideas about ways to maintain the momentum. Their suggestions are listed in Appendix A7.

Participants heartily agreed that the entire conference had been interactive and had modeled the principles of cooperative learning, interaction, and meaningful group activity. There was agreement that future conferences should be held on a regular basis, possibly every one to three years. Participants urged the use of the Internet via Gopher and/or the WorldWide Web in developing an electronic bulletin board or listserve capability. They strongly urged that decision case writers consider publication in refereed journals such as *The Journal of Natural Resources and Life Sciences Education, Hort Technology* and the *Journal of Agribusiness*. The lack of appropriate cases is one of the limiting factors to integrating decision cases into instructional programs for the agricultural and natural resource sciences. The Internet was seen as one way of disseminating information about existing cases and as a mechanism to solicit cases that pertain to specific subject matter.

Conference organizers — The University of Minnesota, Oregon State University, and Michigan State University — unanimously agreed that the conference had been a truly collaborative effort, one that was organized through the use of e-mail and telephone conferences. A commitment was made to continue the collaboration, and other interested persons indicated their desire to join the collaboration. The conference was an excellent example of what happens when collaboration works.

Participants concluded the conference with a renewed commitment to active learning, to the development of additional decision cases that pertain to the disciplines of agriculture and natural resources, to continue networking, and to increase the focus on ways to effectively use decision cases in educational programs.

Part IB
Workshop
Cases

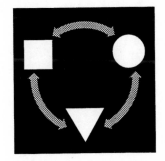

The Worth of a Sparrow

by

R. Kent Crookston,
Melvin J. Stanford,
Steve R. Simmons

THE CASE

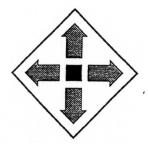

Program for Decision Cases
College of Agriculture
411 Borlaug Hall
University of Minnesota
St. Paul, Minnesota 55108
(612) 624-1211

1
2
3
4
5
6
7
8
9
10
11
12
13
14
15
16
17
18
19
20
21
22
23
24
25
26
27
28
29
30
31
32
33
34
35
36
37
38
39
40
41
42
43
44
45
46
47
48
49
50
51

The Worth of a Sparrow

Who sees with equal eye, as God of old,
A hero perish or a sparrow fall,
Atoms or systems into ruin hurl'd,
And now a bubble burst, and now a world.

Alexander Pope
Essay on Man, I, 11. 87ff.

As Phil Larsen drove past the University's research fields on June 6, 1991, his eyes took in the deep green patches of winter wheat laid out in squares and, beyond them, plots of stout barley and pale green oats. 'A regular smorgasbord;' he thought 'just like Wharton says, one big bird feeder.'

"How'd I get into this mess anyway?" he asked out loud. Phil was Head of the Department of Plant Pathology at the University of Minnesota's agriculture campus in Saint Paul. He had just left a meeting of the Bird Control Committee; he was the Committee chair.

Phil's "mess" had started about a year earlier, on August 1, 1990. On that day he had been heading home for supper via this same route when he noticed a remote telecasting van for a local T.V. station parked in the plots with people standing around one of the bird-cage traps. "I was aware of the Animal Rights Commission's concern about the trapping of birds on campus and realized immediately what was going on," he said. With mixed feelings of responsibility and curiosity, he drove over to the site. His anxiety increased as he discovered that the crew was not just filming; it was telecasting live the lead story for KSTP's Six O'clock Evening News, one of the Twin Cities' major news programs. (Exhibit 1, video segment) .

Although Phil had always known that live-trapping of birds represented a public relations problem, he had never guessed it would come to this. As the prospect of serving as a live-telecast spokesperson for the University on the sensitive issue of humane treatment of animals presented itself before him, Phil became keenly aware of the burden of public accountability that went with his administrative job.

Over the past several years the University had been the focus of a continuing barrage of negative or unflattering news stories. The stories accused the University of environmental insensitivity, harboring reprobate athletes, an inefficient physical plant, etc. The coverage of one 1987 story about alleged misuse of state funds had resulted in the resignation of the University President.

To his relief, Phil was not asked to speak before the cameras that night. Had the telecasters known he was head of the department responsible for bird control, he might not have been so lucky. His involvement with the Animal Rights Coalition (ARC) was certainly not over however; it had just begun.

Phil had joined the University of Minnesota in 1985. When he first learned that his department's responsibilities included controlling birds in the campus cereal research plots, he didn't give the responsibility much thought. "I wondered how we got the job," he said. "I was simply told that Plant Path had always done it. I never did question our control procedure."

The University of Minnesota's research fields were annually plagued by birds which flew in from surrounding urban neighborhoods to feed on the ripening grain. Several control measures had been implemented with varying success depending on the crop and season. Ears of corn were individually bagged with paper as the crop approached maturity; this proved to be costly but highly effective. Birds did not attack soybeans. The small grain plots, on the other hand, had been subjected to very serious bird damage. In addition to a variety of control methods (including

1   scarecrows, noise guns, and scare balloons), traps were placed near the most vulnerable plots.
2   Phil found records indicating that the traps had been used continuously since 1955. In two
3   of the years during that period more than 10,000 birds were trapped, but the average was closer to
4   half that number (Exhibit 2). They were "live" traps; birds were lured inside with pieces of bread
5   where, because of the trap design, they were unable to escape. Water and shade were provided so
6   the birds were contained in relative tranquility until someone came to empty them. "Desirable"
7   birds such as doves and song birds were set free, but "undesirable" ones such as sparrows,
8   starlings and grackles were stuffed into a bag and suffocated.
9   For many years the suffocated birds had been discarded, but beginning in the 1980's they
10 were provided an alternate fate. In the 1970's the University had established a raptor center, which
11 began as a hospital for injured wild birds of prey. By 1990 the Center had become very popular
12 nationally and treated over 500 injured birds, almost half of which were returned to the wild. It
13 had developed into a renowned research and education center and in 1990 had accommodated
14 73,000 school children who came as curious visitors to learn about some of nature's most
15 fascinating creatures.
16   Many of these raptors (kestrels, falcons, etc.) were natural bird hunters; small birds were
17 an essential part of their diet. It was therefore convenient for both the cereal researchers and the
18 Raptor Center researchers to get together. Beginning in the mid-1980's the Center fed all of the
19 birds trapped and not released (about 6,000 each year) to raptors in their care.
20   Over the years, the bird traps had been a subject of controversy. A 1981 letter from one of
21 the University's mathematics professors to the College of Agriculture Dean called for abandonment
22 of bird trapping in the University's agricultural fields. The professor noted how disturbing it
23 would be to one day have University bird trapping featured on the evening T.V. news or in the
24 morning newspaper (Exhibit 3). The Dean acknowledged the potential negative impact of the
25 practice on public support of the University and said he would urge serious consideration of
26 alternative methods of protecting the crops (Exhibit 4).
27   Faculty within Phil's own department questioned the adequacy of the bird control
28 measures. In July of 1987 one of the cereal pathology professors complained that the traps and
29 other methods of control were not successfully protecting his research plots and requested that a
30 committee be formed to determine effective control measures (Exhibit 5).
31   Phil shared the letter with a few interested staff in the Department. One professor
32 responded that the present bird management was expensive and apparently not too effective. He
33 suggested that the research be moved out of the cities "to avoid the social and pest problems
34 currently being experienced." The University held more than 1,000 acres of research land, of
35 quality equal to the campus land, about 20 miles south of the campus.
36   Moving the research out of the cities was felt to be the "last resort" by most of the
37 University cereal researchers who considered the availability of extensive plot areas adjacent to
38 their offices and laboratories to be of "inestimable value." Yet, because of their convenient in-city
39 location, the on-campus plots were annually raided by thousands of birds that nested in the trees of
40 the surrounding suburban yards. The University's cereal plots located out of the city were spared
41 from bird damage, apparently because of their position within a landscape of similar fields and
42 because birds were fewer in number in the country.
43   For the six years prior to the summer of 1990, Phil Larsen had given little thought to the
44 bird control issue. Other issues had kept his thoughts occupied, among them the fact that the Plant
45 Pathology Department had suffered very steep cuts in its education funding because of low
46 undergraduate enrollment.
47   Cereal breeders in the Agronomy Department, plus virtually all of his own faculty and staff
48 that worked in the fields, recommended that the traps and other control methods be maintained.
49 "The traps are highly effective," said Dann Adair, the field plot supervisor who managed them.
50   Then, on August 1, 1990, the mathematics professor's decade-old fears (Exhibit 3) were
51 realized. Both the evening telecast and the next morning's newspaper featured bird killing at the

1   University. On August 2, the morning after the telecast, Mr. Dan Oldre, Vice President of ARC in
2   Minnesota hand delivered a letter to the dean of the College of Agriculture, (Exhibit 6). ARC had
3   conducted a surveillance and investigation and determined that the University had been killing
4   10,000 birds per year by trapping and then suffocating them, 400-500 at a time, in a bag. ARC
5   objected to this cruel wasting of animal life, declared it ineffective in controlling the birds, and
6   insisted that the University "immediately stop the killing of these birds." The letter was copied to
7   several University administrators, and many of the television, radio and newspaper outlets in the
8   Twin Cities. The KSTP telecast had preceded the letter because of a tip given to a KSTP reporter.
9   Both the Minneapolis Star Tribune and the Saint Paul Pioneer Press ran articles about the
10 confrontation between the University and ARC. Phil was quoted in the August 4 Star Tribune as
11 saying "We certainly do not enjoy killing birds, but I think it needs to be said that these plants are
12 extremely valuable and will be useful in the development of high-yielding plants that will feed
13 people worldwide." Oldre's response was "We're not saying their experiments are not valuable,
14 and we're not saying the birds are not a threat, but the method they are using to control the birds is
15 cruel and extreme.... We want them to dismantle the cages and get them out of there and never try
16 that again."
17   Dean Wharton consulted with Phil Larsen and Dann Adair and, on August 6, responded to
18 the Oldre letter. He promised to pursue alternative methods of controlling the birds, and offered to
19 meet with ARC members for a discussion, (Exhibit 7).
20   Up until the August 2 ARC letter, birds caught in University traps were suffocated. On
21 August 7th, the University's newspaper, The Minnesota Daily, ran an article entitled 'Researchers
22 pick CO_2 as bird control choice.' The article reported Dann Adair as saying "This is still a form of
23 suffocation, but it may be less stressful for the birds. We want to do the most humane thing.
24 We'll try it out and experiment with it to see if it's a good way."
25   In the same Daily article, Guy Hodge, staff naturalist for the Humane Society in
26 Washington, D.C. was quoted as saying, "By trapping the birds, you have only reduced the
27 competition for ideal nesting sites. What you end up with is an increased survival rate among the
28 [remaining] birds, thus increasing the population." The article quoted "University officials" as
29 saying, "We have detailed records kept from 1955 saying the traps do undoubtedly make a big
30 difference in reducing the numbers of birds."
31   A meeting between University and ARC representatives was held on Friday, August 10th.
32 On Monday the 13th, the Daily ran an article which reported that the University and the ARC had
33 agreed on the use of bird traps for the rest of the season, but that the ARC would insist on more
34 humane methods next year (Exhibit 8).
35   An August 13 letter from Mary Britton Clouse, an ARC representative, to Dean Wharton
36 reviewed several of the agreements that she understood had been reached during that meeting
37 (Exhibit 9). On August 20, Phil Larsen wrote to Dean Wharton with reactions to several of Mary
38 Britton Clouse's impressions about agreements reached at the August 10 meeting (Exhibit 10). An
39 August 22 letter from Dean Wharton to Ms. Clouse clarified the University's understandings.
40 Dean Wharton was careful to point out that although the University would explore additional and
41 alternative methods for controlling the birds, it had not agreed to stop trapping (Exhibit 11).
42   About this time, Phil began receiving some supportive comments from members of the
43 community. A telephone message taken by his secretary read "Any time you kill a grackle or
44 sparrow, we bird watchers really appreciate it." A woman from Saint Paul sent a $10.00
45 contribution in support of the bird control program with a note that said "Please accept this small
46 donation to help you smother sparrows and starlings who attack your crop next year."
47   A campus employee sent Phil an article that appeared in a weekly paper called "Minnesota
48 Outdoor News." The article was about an organization called Putting People First, a group of
49 concerned citizens who objected to being intimidated by "animal right activists." Accompanying
50 the article was a petition to the Congress of the United States requesting that votes by Congress
51 reflect the majority of American citizens, not just "a minority of vocal extremists." There was a

1 place to sign the petition, and instructions for joining the organization and/or for sending money to
2 support it (Exhibit 12).
3   It was in this climate of diverse viewpoints that Phil pondered his course of action between
4 the 1990 and 1991 growing seasons. "It seems to me that this grain is important enough to justify
5 the sacrifice of these birds," he said to colleagues. "There are world hunger and economic
6 implications. We have to consider our right to prioritize research of value to humans over creature
7 life. And there's another, kinda far out issue," Phil offered. "Do plants have a right to protection?
8 I've heard it asked several times."
9   On September 7, 1990, Phil met with several interested University faculty and field staff to
10 consider the issues. Minutes from the meeting included the following statements: "The
11 effectiveness of our current bird control program was discussed and it was agreed that the current
12 program was effective and needs to be continued.... Phil Larsen appointed a committee, for which
13 he will serve as chairperson, to develop and establish an integrated plan for bird control on the
14 Saint Paul plots that includes coordination of all individual projects so that everyone knows what
15 our overall objectives are."
16   The first Bird Control Committee meeting was held on November 6, 1990. There was a
17 discussion of present techniques being used for control including trapping, use of ribbons and
18 balloons, alternate feeding sites, and raptors (a falconer had been employed from 1982 to 1990 to
19 visit the plots during periods of heavy bird feeding. He not only helped collect trapped birds for
20 the Raptor Center, but brought live raptors with him which he flew over the plots to scare the
21 birds). Following a discussion of the merits and shortcomings of current approaches, committee
22 members were assigned to check out other control options including hawk-shaped balloons,
23 recorded distress calls, and applications of hot pepper oils to the grain.
24   The next Bird Control Committee meeting was held on December 10, 1990. Progress
25 reports were given on assignments from the previous meeting, and there was a discussion about
26 sending representatives to the upcoming Great Plains Animal Damage Control meetings to be held
27 in Nebraska.
28   At the March, 1991, Bird Control Committee meeting an update on alternate control
29 methods suggested several promising options. John Arent (the falconer) and colleagues reported
30 on a study they had initiated to evaluate various sprayed repellents. Phil reported a meeting with
31 Mary Britton Clouse during which Mary provided additional bird control information and informed
32 him that the Humane Society of the United States provided small grants (approximately $3,000 to
33 $5,000) for research. The College of Agriculture's Public Relations Officer attended and
34 discussed proper handling of public interactions.
35   At an April 25, 1991, meeting of the Committee Dann Adair and John Arent gave a report
36 on the Animal Control Conference they had attended in Lincoln. They indicated that there were
37 lots of ideas that "someone said" worked, but nothing of particular promise. The group recognized
38 the need for a statement for the media in anticipation of objections to the bird control program for
39 the 1991 season. It was agreed that Phil Larsen and Dann Adair would prepare a one-page
40 document providing an overview of the College's bird control program to be posted on the bird
41 traps and anywhere else that might be appropriate for the purposes of communicating the issues.
42   What to do in the case of a hostile confrontation was discussed during the April 25
43 meeting. Phil observed that "both Dann and John know that this is a potentially very delicate
44 issue, and that they will be under intense scrutiny." This led to the question of abandoning
45 trapping, but "capitulation" was felt to be out of the question by most of the committee. The
46 possibility of trapping and re-releasing the birds was suggested (re-releasing at night, or at some
47 distant location), but was considered impractical. Toward the end of the meeting, one of the cereal
48 breeders said, "Right now let's not confuse the process. Let's go forward; follow the KISS
49 principle." Phil knew this meant Keep It Simple Stupid.
50   On May 29, 1991, the Committee met again to make specific decisions for implementing a
51 bird control program for the coming year. It was decided to employ traps as early as June 15 in

1   order to protect the winter wheat that would be forming grain. Various assignments were made
2   including an evaluation of sprayed grape juice concentrates (reported to repel birds), sprayed hot
3   pepper sauce, balloons, ribbons, loud speakers with distress calls, etc. The plan was to vary the
4   approach throughout the season to prevent birds from becoming accustomed to any one control
5   tactic. A June 6 meeting was planned for all faculty and technicians from participating projects to
6   inform them about, and to solicit their participation in, the control program.
7   There was a small crowd at the June 6 meeting. Larry Hood, Assistant Director of the Law
8   Enforcement Division of the U.S. Fish and Wildlife Service, was there. Phil distributed a handout
9   containing information about the bird damage management program for the 1991 season. It noted
10   that birds would be trapped if necessary (Exhibit 13).
11   A few questions were asked, the history of the University/ARC interactions was briefly
12   reviewed, and then Larry Hood initiated an interesting dialogue.
13   "Think about how this looks folks," he began. "If you don't let the momma bird go back
14   to her nest, the young back in the nest will suffer. The 'humaneiacs' will nail you if they find one
15   dead robin, bluebird, or dove in your traps. And you will be extremely vulnerable. You will have
16   violated the migratory bird laws. Those laws have been here since 1918."
17   "You gotta realize that they're on us," Larry continued. "They ask me: 'Why are you not
18   enforcing the law?'"
19   "If one protected bird is in your trap you're in violation," Larry warned. "Could be a
20   twenty thousand dollar penalty, maybe a hundred thousand. The problem with trapping is, you
21   attract them. Mourning doves will give you fits."
22   "What's to stop them from planting a protected bird in our traps?" asked one professor.
23   "How can we avoid being framed?"
24   "You can't," answered Larry.
25   "So why don't we stop trapping?" asked the College's public relations officer.
26   Dann Adair's response came quickly: "It's clearly valuable to trap."
27   "How do we know that?" asked another professor. "Does trapping really help, or is it a
28   matter of revenge? Isn't it possible the baited traps actually attract birds? Do we have any data on
29   the effect of traps and actual damage to the grain?"
30   "Tell us how to conduct the study and we'll do it," responded Dann Adair.
31   Phil was as confused and concerned as any in the group by what he was hearing. It was
32   obvious that some felt the trapping should be stopped. He looked at his watch. It was time to
33   leave for an off-campus meeting. He asked Dann Adair to field any remaining questions and
34   excused himself.
35   And so it was that on the morning of June 6, 1991, Phil Larsen was again driving past the
36   cereal plots asking out loud, "How'd I get into this mess anyway?" A colleague was with Phil in
37   the car; he had also been in attendance at the June 6 meeting. Phil did most of the talking.
38   "We wouldn't be into this if it wasn't for the ARC," he said. "But I've got to give them
39   credit. We're all a lot more conscious and sensitive about the way we view the life of a creature.
40   We've done a lot of things to get our act together, things we would never have tried on our own.
41   But, I'd like to get us to a place where we say enough is enough. We need to be ethically
42   responsible, but we can't go overboard. They have taken the issue and exploited the media to
43   influence public opinion in their favor. Will we really serve the public if we give in to them?"
44   "And there's another thing about all this," Phil continued. "I'm concerned about
45   vandalism. I've heard they have the potential to act as terrorists. I've never feared for my life, but
46   I've definitely been concerned about vandalism of the plots."
47   Phil looked at this colleague. "What's the right thing to do?" he asked.

The Worth of a Sparrow

by

R. Kent Crookston,
Melvin J. Stanford,
Steve R. Simmons

EXHIBITS

Program for Decision Cases
College of Agriculture
411 Borlaug Hall
University of Minnesota
St. Paul, Minnesota 55108
(612) 624-1211

The Worth of a Sparrow
Exhibit 1

A video recording of the lead story for KSTP's 6 o'clock news,
August 1, 1990

The Worth of a Sparrow
Exhibit 2

BIRD TRAPPING - UNIVERSITY OF MINNESOTA, AGRICULTURAL EXPERIMENT STATION, ST. PAUL
KINDS AND NUMBERS OF BIRDS TRAPPED

YEAR	AVE. TRAPP-ED PER DAY	SPARROWS	BLACK-BIRDS	STARLINGS	GRACKLES	COW-BIRDS	TOTAL
1955[1]	40	7,551	137	1,348	159	-	9,195
56[1]	24	7,128	118	1,459	137	-	8,842
57	108	6,129	1,159	1,796	244	298	9,626
58	121	8,662	285	1,180	264	9	10,400
59	78	4,128	257	619	106	137	5,247
1960	106	6,305	581	1,086	163	-	8,135
61	119	6,135	332	739	12	164	7,382
62	101	6,172	986	1,601	466	128	9,353
63	101	6,811	229	314	113	15	7,482
64	120	7,521	65	840	25	49	8,500
65	87	3,698	1,630	1,110	29	401	6,868
66	70	3,960	366	9	117	11	4,463
67	130	6,959	1,193	575	251	660	9,638
68	120	4,553	708	407	390	541	6,599
69	84	2,086	598	341	72	688	3,785
1970							(Rec. Missing)
71	96	2,317	1,468	400	31	91	4,307
72	93	3,647	626	124	277	440	5,114
73	52	2,105	856	112	334	-	3,407
74	95	3,156	346	1,989	248	406	6,145
75	-	-	-	-	-	-	-
76	83	2,829	264	905	147	607	4,752
77	201	4,907	1,149	887	446	472	9,846
78	-	-	-	-	-	-	-
79	77	3,752	229	576	17	315	4,989
1980	91	3,843	502	1,104	325	714	6,488
81	99	4,701	617	272	87	337	6,014
82[2,3]	-	522	95	43	113	-	773
83	-	346	52	218	119	-	735
84	-	-	-	-	-	-	-
85	-	1,824	684	912	228	-	3,648
86	-	3,121	1,568	1,235	398	-	6,322
87	-	1,038	156	804	207	-	2,205
88	-	2,731	1,758	1,698	-	-	6,187
89	-	3,977	2,943	3,460	-	-	10,380
90	-	3,396	395	2,160	-	-	5,951

[1]Trapping by Twin City Pigeon Eliminating Company.
[2]Trapping contracted with John Arent through 1990.
[3]1982 - 1990 - Cowbirds included with blackbirds.

The Worth of a Sparrow
Exhibit 3

UNIVERSITY OF MINNESOTA   School of Mathematics
TWIN CITIES   127 Vincent Hall
                            206 Church Street S.E.
                            Minneapolis. Minnesota 55455

July 21, 1981

Dean James Tammen
College of Agriculture
University of Minnesota

Dear Dean Tammen:

I urge you to consider seriously the abandonment of the policy of trapping birds
in the University agricultural fields. I consider this practice as having the
potential to create a serious public relations problem during a difficult financial period for the University.

As a resident of one of the neighborhoods surrounding the University fields, I
was able to observe the reaction of some of my neighbors to the introduction of
bird traps in the field near the intersection of Snelling and Roselawn. The
traps were placed close to a public path in clear view of pedestrians and
bicyclists. It is almost impossible to avoid seeing the distressed birds from
the path. Such a sight is disturbing to many persons, particularly when they
learn from the students in the field that the birds are being destroyed.
Although these birds are perceived as pests by the University staff conducting
experiments in the field, they are perceived as attractive representatives of
nature by many of the local residents, who feed them during the winter and
encourage them to remain in their yards and gardens.

It is my impression that the residents of the area are very friendly toward the
University in general and the agricultural experiments in particular. However,
this summer's trapping follows on the heels of the accidental poisoning last
summer of a large flock of birds in the same field. Some of the residents are
beginning to perceive the University as a heartless killer of wildlife. I would
hate to see on the evening news television pictures of panicked birds fluttering
wildly in University bird traps or to read in the morning paper an article about
bird killing at the University.

I urge you to find alternate methods for protecting the experiments. Perhaps
placing nets over certain fields during critical stages of the experiments would
provide more effective protection. The additional cost would be well spent,
producing better data and a better public image for the University.

Thank you for your attention to this matter.

Sincerely,

Richard McGehee
Professor

cc. Dean William Hueg, Jr., Institute of Agriculture, Forestry, & Home Economics
    Professor Kenneth Keller, Vice President for Academic Affairs

Office of the Dean

The Worth of a Sparrow
Exhibit 4

UNIVERSITY OF MINNESOTA
TWIN CITIES

College of Agriculture
277 Coffey Hall
1420 Eckles Avenue
St. Paul, Minnesota 55108

(612) 373-0921

July 24, 1981

Dr. Richard McGehee
Professor
127 Vincent Hall
Minneapolis Campus

Dear Dr. McGehee:

I am writing to thank you for your thoughtful letter of July 21,
1981 in which you express your genuine concern about efforts to reduce the
depredation of experimental crops by birds and the adverse impact that
this may have upon publics supportive of the University in general and
of agricultural research in particular. Because the experimental plots
form a part of the Agricultural Experiment Station, a unit of the Institute
of Agriculture, Forestry and Home Economics administered separately from
the College of Agriculture, I have brought your letter to the attention
of the Director, Dr. Richard J. Sauer. Dr. Sauer informed me that Dr.
David W. French, Head of the Department of Plant Pathology, is the person
responsible for dealing with this matter. For this reason I am forwarding
your letter to Dr. Sauer and to Dr. French asking that they give immediate
attention to the questions you raise. I know that Dr. French and/or
Dr. Sauer will be pleased to discuss this difficult matter with you and
to give serious consideration to alternative methods of protecting
critically important experimental plantings from birds consistant with
good wildlife management practices.

Sincerely yours,

James F. Tammen
Dean

JFT:map

cc: Dr. R. J. Sauer
    Dr. D. W. French
    Dr. W. F. Hueg
    Dr. K. Keller

The Worth of a Sparrow
Exhibit 5

UNIVERSITY OF MINNESOTA   Department of Plant Pathology
TWIN CITIES   495 Borlaug Hall
                            1991 Upper Buford Circle
                            St. Paul, Minnesota 55108

                            (612) 625-8200

July 29, 1987

Dear Phil:

The bird problem in small grain research plots at St. Paul is a serious
problem. This problem has been on-going for many years, but this year it was
almost beyond toleration. It affects my field research on kernel
discoloration and scab, Don McVey's rust research, as well as that of Agronomy
Department.

Present efforts to reduce the size of the bird population are a failure. I
believe this failure results from:

1)   Complex habits of the birds and favorable habitat on campus and in
      surrounding neighborhoods.
2)   A winter in 1986-87 that was mild and hence favorable for bird survival.
3)   A lack of real concern by the persons who, the Project leaders, have
      thought have responsibility for control.
4)   Lack of an effective plan for control that everyone thinks will work.

To correct the situation I suggest a committee be organized to consider the
problem and to take action.

The committee should be made up of interested persons, persons who have a
vested interest in control. Perhaps the following would be adequate:

1)   Department Head (you and/or Orvin)
2)   Roy Wilcoxson
3)   Don McVey
4)   Deon Stuthman
5)   Bob Busch
6)   David Davis
7)   Dan Adair
8)   A qualified ornithologist who knows how to control birds or at least
      knows their habits.

This committee should:

1)   Consider the problem and whether the population can be reduced.
2)   Consider methodology to reduce the population if reduction is possible.
3)   Identify persons responsible for carrying out the methods as well as the
      persons for supervision. At present, responsibilities for both action
      and supervision is so diffuse that no one really seems to know their
      duty.
4)   Seek resources adequate to carry out the methods agreed effective.
5)   Develop an educational program for persons in the neighborhoods about
      campus to inform them and to request their cooperation.

This committee should be organized soon because they may wish to initiate
activities this winter if they decide the problem can be solved.

Roy K. Wilcoxson

ANIMAL RIGHTS COALITION. INC.

August 2, 1990

Dr. Keith Wharton
Dean, College of Agriculture
University of Minnesota
277 Coffey Hall
1420 Eckles Avenue
St. Paul, MN 55108

Dear Dr. Wharton:

The Animal Rights Coalition was alerted by a concerned person two weeks ago that the University is trapping and killing large numbers of birds in the experimental crop fields on the Saint Paul campus. In our own surveillance and investigation we have learned that:

- Birds are lured to the traps by bait placed in them, and the number of birds in these large traps has been observed to be at least 50 at times.

- The reported method of killing the birds was suffocation, by placing them, 400-500 at a time, in a bag.

- Approximately 500 birds are killed every day at the St. Paul campus; 10,000 were killed in a three month period.

We object to this for the following reasons:

- It is a waste of animal life.

- It is ineffective. Trapping and killing birds will not permanently reduce the bird population in the area. Even *temporarily* it will have no more than a minimal effect.

- The trapping and suffocation of these birds is cruel in the extreme.

P.O. BOX 20315 • BLOOMINGTON, MINNESOTA 55420 • (612) 822-6161 • (612) 888-0288

page 2   The Worth of a Sparrow
Dr. Keith Wharton   Exhibit 6, Page 2 of 2

For these reasons, **we insist that the University immediately stop the killing of these birds.**

If you wish to discuss this matter, please contact me at 222-5537 or 870-5868.

Dan Oldre
Vice President
Animal Rights Coalition

cc:
Dr. Phil Larson, Head   KSTP-TV
Plant Pathology Dept.   WCCO-TV
University of Minnsota   KMSP-TV
495 Borlaug Hall   KARE-TV
1991 Upper Buford Circle   WCCO-Radio
St. Paul, MN 55108   Minnesota Public Radio
                                         KFAI-Radio
President Nils Hasselmo   Associated Press
University of Minnesota   Minnesota Daily
202 Morrill Hall   Saint Paul Pioneer Press
100 Church Street SE   Star Tribune
Minneapolis, MN 55455   Twin Cities Reader

Kent Crookston, Head
Agronomy and Plant Genetics
University of Minnesota
411 Borlaug Hall
1991 Upper Buford Circle
St. Paul, MN 55108

C. Eugene Allen, Director
Agriculture Experiment Station
University of Minnesota
220 Coffey Hall
1420 Eckles Avenue
St. Paul, MN 55108

Office of the Dean

UNIVERSITY OF MINNESOTA   College of Agriculture
TWIN CITIES   277 Coffey Hall
                            1420 Eckles Avenue
                            St. Paul, Minnesota 55108-1030

August 6, 1990

Mr. Dan Oldre
Vice President, Animal Rights Coalition
P.O. Box 20315
Bloomington, Minnesota 55420

Dear Mr. Oldre:

I am replying to your letter of August 2 to me concerning the
trapping and killing of birds in the experimental crop fields on the
Saint Paul Campus of the University of Minnesota.

I have talked with Dr. Philip Larsen, Head of the Department of Plant
Pathology, and Mr. Dann Adair, Plant Pathology Research Plot
Coodinator. They have assured me, as they conveyed to you earlier,
that the purpose of the bird control program is to protect the
experimental plantings that have been initiated for development of
new crop varieties. The seed produced on these plots is extremely
valuable because it represents the product of genetic crosses that
have been carried out over a period of years. With each plant
selection there may be only a few seeds produced, so it is essential
that they be protected. The new plant varieties produced from this
research will be used to increase and improve food production and
quality on a worldwide scale, thus having a dramatic impact on
reducing world hunger. We believe the benefits of this plant
breeding research are important enough that we must protect these
plants from destruction by pest birds.

The research plots on the Saint Paul Campus occupy approximately
100 acres. The bird control program, which usually lasts only for
about 6-8 weeks, from late June until the middle of August, is a
comprehensive one that involves several approaches. Included are:

Mr. Dan Oldre
August 6, 1990
Page 2

* Planting border strips of grain around the plots to encourage the birds to feed there rather than on the research plots.

* Attaching plastic strings to poles around the plots. These strings vibrate in the wind and produce a noise that frightens the birds.

* Suspending large, brightly-colored balloons with metallic designs (scare eyes) on poles throughout the plots.

* Working with a licensed falconer who uses trained hawks to frighten the birds away from the plots.

* Attempting to encourage the nesting of sparrow hawks in the plot area to help control the birds.

* Trapping and destroying the "pest" birds.

We have considered placing nets over the plots, as you suggested, but have determined that this would be both impractical and ineffective.

The birds are lured away from the research plots to eight large cages baited with bread and water scattered throughout the plots. The trapped birds are held in these traps, with food and water, until they are removed at the end of the day. At that time all songbirds are released, and the "pest" birds -- blackbirds, starlings, grackles, and English sparrows -- are taken to another location and humanely killed and given to the University Raptor Center for feed for the birds that are housed there.

Although the number of birds trapped daily varies widely, our records show that approximately 150-200 birds total will be trapped daily on the average. The average number trapped per year is usually about 5,000.

Mr. Dan Oldre
August 6, 1990
Page 3

While we believe that the method we have been using to destroy the pest birds is a humane one, we will, for the remainder of this season, change to the use of carbon dioxide to destroy them. The birds will be collected in a cloth bag and kept in a cool place until they are destroyed with a high concentration of carbon dioxide.

We do not like having to destroy birds, and will continue to explore, evaluate, and improve alternative approaches to controlling them. As I stated earlier, however, the research on these plots is extremely important and must be continued.

As I told you when we talked on the telephone last Friday afternoon, I and others from the University will be pleased to meet with you and others from your organization to discuss this situation. Please let me know if you would like to meet, and we can arrange a time.

Sincerely,

Keith Wharton
Acting Dean

Copy: President Nils Hasselmo
          Vice President C. Eugene Allen
          Dr. R. Kent Crookston
          Dr. Philip Larsen
          Dann Adair
          Jeff Wilson
          Nina Brook
          Mary Ann Grossman
          Erin Rasmussen
          Jeff Crilley
          Lori Kroontje

The Worth of a Sparrow
Exhibit 8

U, ARC agree on use of bird traps

UN DAILY
9/13/90

By **Lori Kroontje**
Staff Reporter

The College of Agriculture will continue to use bird traps for the rest of the year, it announced Friday at a meeting with the Animal Rights Coalition.

The coalition met with college representatives to discuss use of the bird traps, which catch birds disrupting the college's experimental grain field.

The birds caught — excluding song birds — are killed with carbon dioxide and fed to recovering falcons and hawks in the Raptor Center.

Since only one week remains until harvest, college representatives said it was too late to start a new method of trapping the birds.

But coalition members said next year they will insist the University use a more humane method of deterring birds from its research.

At the request of the coalition, researchers began suffocating the trapped birds with carbon dioxide last week rather than suffocating them in plastic bags. But ARC members said they won't be satisfied until the killing stops altogether.

"I'm pleased that they at least agreed to try our suggestions," said coalition member Kathy Laszlow. "But I feel like they want us to do all the

See Birds page 3

Birds *from 1*

research. It still remains our responsibility to come up with a better solution."

At Friday's meeting, the coalition suggested two alternatives for deterring the birds — covering the plots with netting and transporting the birds to another area.

Both suggestions, however, have drawbacks.

Covering the 100-acre experimental station with a net would cost the University at least $120,000, said Harold Stump, owner of R & D Equipment, a bird-net company in Sioux City, Iowa.

Also, the netting would not withstand sun exposure, even with ultraviolet protectants, and would need to be replaced every five to six years, Stump said.

The College of Agriculture, now undergoing retrenchment and facing a $250,000 cutback in its budget next year, could not afford such expenses, said Phil Larson, head of Plant Pathology.

Coalition members volunteered to help with their second solution — transporting the birds to another location.

University officials agreed to this, but as coalition members discovered later in the day, the plan was not as simple as it sounded.

After calling the Department of

Photo/**Mark Trockman**

Falcons like this one are employed to spook birds at experimental plots on the St. Paul campus.

Natural Resources, coalition member Mary Britton-Clouse discovered transporting birds to the country could upset the existing ecological system.

In addition, Twin Cities neighborhoods may not want the birds — sparrows, starlings and blackbirds — because they are dirty and spread disease, said Dan Adair, director of the Plant Pathology greenhouse.

The fact that the experimental station is located in the center of a large city further complicates the problem.

Grain at the University's Rosemount Agricultural Experiment Station, for example, is not plagued by birds. Rosemount, outside the metropolitan area, has a greater food supply for the birds than the Twin Cities, said College of Agriculture Dean Keith Wharton.

"The St. Paul campus is one big bird feeder," he said.

In addition, the pest birds nest in areas such as house eaves, which makes them more prominent in the city, Wharton said.

University officials and coalition members said they were satisfied with the way the meeting recognized the importance of research as well as animal rights.

"I guarantee I will look very carefully into these suggestions," Wharton said. "I can't guarantee they'll work, but I will look into them."

ANIMAL RIGHTS COALITION, INC.

August 13, 1990

Dr. Keith Wharton
Dean, College of Agriculture
University of Minnesota
277 Coffey Hall
1420 Eckles Avenue
St. Paul, MN 55108

Dear Dr. Wharton,

Thank you for providing the Animal Rights Coalition representatives with the opportunity to meet with you and other University members regarding the trapping and killing of birds in the experimental crop fields. We feel that a great deal was accomplished in clarifying the issues and look forward to a resolution that will satisfy all concerned.

I would like to review the following conclusions of that meeting. It was agreed that:

·The Department of Plant Pathology commits itself to researching bird deterrent methods with the ultimate goal of eliminating the current method of trapping and killing by next season of late June to mid-August, 1991.

·The Department of Plant Pathology will consult with the attatched names and organizations, as well as any other resources at the department's disposal, for information and advice regarding alternatives. Please provide ARC with progress reports on or by December 1, March 1 and June 1.

·Several diverse alternative methods will be correctly applied and given an adequate length of time, and monitored and documented to assess their effectiveness.

The Worth of a Sparrow
Exhibit 9, Page 2 of 3

Page 2

·ARC will be provided with documentation of previous bird statistics from 1955 to present, and will be provided with new statistics at the end of the 1991 season.

·The Department of Plant Pathology will work with ARC member Gary Messerich during the remainder of the 1990 season to determine a means by which birds trapped can be released.

Please direct all questions, comments and correspondence to me. Again, please accept out sincere thanks for your cooperation in this matter.

Respectfully,

Mary Britton Clouse
Animal Rights Coalition
2023 Lowry Avenue N
Minneapolis, MN 55411
(612) 626-0303

cc.- President Nils Hasselmo, Vice President C. Eugene Allen, Dr. R. Kent Crookston, Dr. Philip Larsen, Dann Adair, KSTP-TV, WCCO-TV,KMSP-TV, KARE-TV, WCCO-Radio, Minnesota Public Radio, KFAI- Radio, Associated Press, Minnesota Daily, St. Paul Pioneer Press, Star Tribune, Twin Cities Reader

Page 3

Bird Control Resources

The Humane Society Of the United States   (202) 452-1100
Guy Hodge, Naturalist
2100 L Street NW
Washington, DC 20037

International Alliance for Sustainable Agriculture
Terry Gips, Executive Director   (612) 331-1099
Newman Center, University of Minnesota
1701 University Avenue SE
Minneapolis, MN 55414

Professor Ronald Johnson   (402) 472-6823
202 Natural Resources Hall
University of Nebraska
Lincoln, Nebraska 68583

David Tressemer (Bird Control Consultant)   (303) 449-0486
Boulder Colorado
(has citations on USDA Department of Agricultural Ornithology,
c.1880-1930)

The Worth of a Sparrow
Exhibit 10

UNIVERSITY OF MINNESOTA   Department of Plant Pathology
TWIN CITIES   495 Borlaug Hall
                                 1991 Upper Buford Circle
                                 St. Paul, Minnesota 55108

                                 (612) 625-8200
                                 FAX: (612) 625-9728

August 20, 1990

TO: Keith Wharton

FROM: Phil Larsen   *Phil*

RE: **AUGUST 13 ARC LETTER TO YOU**

I wanted to give you my reaction to several of the points made by Mary Britton Clouse and her impression about what was agreed upon at our meeting with ARC representatives. First, we did commit ourselves to researching alternative bird deterrent methods but we did not agree to an ultimate goal of eliminating the current method of trapping and killing by next season. I would certainly admit that it would be desirable to be able to accomplish those goals but it may not be possible.

Regarding the comment that we agreed to consult the attached names and organizations, etc., - we intend to consult with any recognized authority on bird control or bird behavior that we choose. However, Terry Gips (Sp?) would not be considered to be in this category. The other three individuals mentioned certainly seem to be good resources based upon their titles, and we will probably contact them, but I don't recall making any promises or obligations that we should contact the people that they chose. We have already contacted Professor Ronald Johnson at the University of Nebraska on one occasion and plan to contact him again, however.

On Page 2, Ms. Britton Clouse indicated that ARC will be provided with previous bird trapping statistics from 1955 to the present and with new statistics at the end of the 1991 season. Although I didn't take notes on this point, I believe I do recall that we would be agreeable to providing this information and will do so.

I have spoken with Gary Messerich on at least two occasions since our August 10 meeting concerning the possibility of transporting the birds to another release site. In his telephone conversations with me, Gary has indicated that he is beginning to see some of the difficulties associated with transporting the birds to another site. He has alternatively suggested that we try releasing the birds at night after trapping them during the day. The rationale being birds only feed during the day and that they would return to their nesting site in the evening. According to Jim Kitts, this approach would quickly train the birds to avoid the traps. It may be worth a try, however, we are at the stage where most of the grain has been harvested and we will be concluding our bird control for this season. Within the next month, I expect to convene a meeting of individuals associated with bird control on the St. Paul Campus including those scientists whose work is directly affected by the bird control. We would also include Jim Kitts and any other individuals who are knowledgeable in the area of bird control and bird behavior on the St. Paul Campus. In addition, we will be calling some of the experts that have been suggested in this letter as I mentioned above. The purpose of the meeting will be to review our bird control program and to see how it might be improved. We are committed to having a bird control program in place that is as humane as possible, but still effective in protecting our experimental plots.

PL:aa

c: Dann Adair

Office of the Dean

UNIVERSITY OF MINNESOTA   College of Agriculture
TWIN CITIES   277 Coffey Hall
                                 1420 Eckles Avenue
                                 St. Paul. Minnesota 55108-1030

August 22, 1990

Ms. Mary Britton Clouse
Animal Rights Coalition
2023 Lowry Avenue North
Minneapolis, Minnesota 55411

Dear Ms. Britton Clouse:

I am pleased that Dann Adair, Richard Jones, Phil Larsen, Dani O'Reilly, and I were able to meet with you and the others from the Animal Rights Coalition (ARC) on Friday, August 10 to discuss the bird control program in the crops research plots on the Saint Paul Campus. I thought this was a productive meeting in that additional information was provided by both the ARC and the University, that all agreed on the importance of the crops research and the necessity to control the damage done by the birds, and that some agreements on future actions were made.

I have read your letter of August 13 to me, and want to respond so there will be no misunderstanding of what these agreements were.

* The ARC and the University agreed that the crops research is important, that it must be continued, that birds do damage the research plots, and that this damage must be controlled.

* The University stated that the bird control measures currently being used would continue throughout the remainder of this season, but that between the end of this season and the start of the control period in 1991 the University will explore additional and alternative methods for controlling the bird damage. This does not mean that the University agrees to change its current program. It means that we will carefully explore valid, reliable, effective, feasible, and humane alternatives, and if one or more can be found that will meet our requirements, they will be adopted.

Ms. Mary Britton Clouse
August 22, 1990
Page 2

* The University is willing to work with ARC during the remainder
of the current season to transport the trapped birds to a release site
provided that an acceptable site can be found that meets all
community and state regulations.

* The University will make available to ARC the bird trapping
statistics from the 1950's to the present.

While we did not agree to provide ARC with periodic progress
reports, we will be happy to keep you informed on the alternative
approaches that are being explored, and will inform you prior to the
beginning of the 1991 control season of the methods that will be
used.

Sincerely,

Keith Wharton
Acting Dean

Copy: Dann Adair
         Richard Jones
         Phil Larsen
         Dani O'Reilly

Petition to the Congress of the United States

WHEREAS, animal "rights" activists openly proclaim that intimidation, arson, property destruction, burglary, and theft are acceptable crimes when they are used as forms of civil protest to promote the interest of animals over people; and

WHEREAS, animal "rights" terrorists have claimed responsibility for more than $10 million in damages to medical research laboratories in the United States and animal advocates prompt the mistaken belief that alternative methods of research exist which will eliminate the need for research on animals; and

WHEREAS, businesses using animals for food, research, entertainment, and clothing are being harassed, their goods are being destroyed, their stores are being vandalized and bombed by pro-animal protesters trying to close down their industries, and their customers are being intimidated; and

WHEREAS, persons claiming that animals are equivalent to human beings are committed to abolishing the use of animals in science, the total elimination of sport hunting and trapping, the total dissolution of any commercial use of animals, and the elimination of the millions of jobs associated with these activities;

NOW, THEREFORE BE IT RESOLVED, that the signers, as Citizens of the United States, do hereby petition the Congress of the United States as follows:

To oppose those who would put millions of Americans out of work by closing any business or industry that entails the use of animals; and

To enact legislation which will combat intimidation, terrorism, and violence as used by animal activists and protesters; and

To continue funding medical research and other testing which uses animals responsibly;

And generally, to elicit the views of *all* their constituents so votes by Congress will reflect the majority of American citizens, not just a minority of vocal extremists.

Signature: _____ Print Name: _____

Address: _____

The Worth of a Sparrow
Exhibit 12, Page 2 of 3

_____We will detach prior to submitting your Petition to Congress _____

TO: Putting People First
        Suite 310-A
        4401 Connecticut Avenue, N.W.
        Washington, DC 20008-2302

Yes, I want to stand up for the interests of people and fight the intimidation tactics of animal "rights" extremists.

☐   Please enroll me in Putting People First. Enclosed is my check for:

    ☐ $15 (Membership dues)   ☐ $25   ☐ $50   ☐ $100

    ☐ $500   ☐ $1,000   ☐ Other

Please make your check payable to *Putting People First.*

☐   To help strengthen our organization's fight against animal extremists and to alert Congress of our just cause, I want to send copies of this Petition to my friends, associates, and clients. Please send _____ additional Petitions.

☐   I want to help start a local chapter of Putting People First. Please contact me. My telephone number is: Area code _____ Telephone _____

☐   No, I can't offer $15 right now, but I have returned my signed Petition and I am behind your efforts 100%

Please Print:

    Name:_____

    Address:_____

    City: _____State: _____Zip Code: _____

Distributed by Putting People First, a nonprofit tax exempt organization.
Contributions to Putting People First are not tax deductible.

1407-2562-2

The Worth of a Sparrow
Exhibit 12, Page 3 of 3

Putting People First is a group of concerned citizens from all walks of life who object to being intimidated by so-called "animal rights" activists. Our organization opposes the terrorists who proclaim that physical harassment, arson, property destruction, burglary, and theft are "acceptable crimes" when used as civil protest to promote the interests of animals over people.

Businesses using animals for food, research, entertainment, and clothing are being harassed, their goods are being destroyed, their stores are being vandalized by pro-animal protesters, and their customers are being intimidated. Pro-animal extremists have claimed responsibility for more than $10 million in arson damages to medical research laboratories in the United States, and animal advocates promote the mistaken belief that alternative experiments exist which will eliminate the need for research on animals.

No ethical person would disagree that mankind must treat all living creatures humanely. In most instances, humans and animals can coexist without any conflict of interest. But from time to time the interests of animals and those of mankind may cross. In the cases of conflict between man and animal, we believe that the interests of people must take precedence over those of animals.

Putting People First is a tax-exempt grassroots organization, with headquarters in Washington, D. C. and local chapters all over America. It is a registered lobby with two registered lobbyists, and sponsors a political action committee. *Putting People First* tracks legislative proposals at both the federal and state levels, and by means of a Legislative Alert system calls on its members to write letters to Congress or local governments to support or oppose bills when key votes are scheduled.

Putting People First does not confine itself to legislative action, however. It files public interest lawsuits in federal and state courts and with federal regulatory agencies to stop animal activists whenever their tactics cross the line to go outside the law. We are tracking media coverage of animal protesters with our in-house clipping service, and we write letters to editors and appear on talk shows to keep a balanced presentation before the American people.

And when animal extremists take on individual researchers and businessmen with their Big Lie tactics and enormous media blitz expenditures, designed to overwhelm their targets into submission, *Putting People First* will stand up and defend these individuals not only with our words but also with action.

Putting People First does not represent any industry or special interest. We represent the average American who drinks milk and eats meat, benefits from medical research, wears leather, wool and fur, hunts and fishes, owns a pet, and goes to zoos. We respect the views of others and support their right to hold those views. But animal activitists would impose their ideology on everybody, without exception, and suppress any values that don't match their own. Our views are equally defensible and we demand the right to hold our beliefs in peace.

4401 Connecticut Avenue N.W., Suite 310-A, Washington D.C. 20008-2302 (202) 364-7277

BIRD DAMAGE MANAGEMENT PROGRAM
ST. PAUL EXPERIMENTAL PLOTS
UNIVERSITY OF MINNESOTA
ST. PAUL

We believe the small grains experimental plots on the St. Paul Campus are extremely important to the livelihood of those who produce these crops in Minnesota and throughout the world. The genetic backgrounds of the plants grown in these plots have been developed over many years and provide a vital link to their genetic improvement. Consequently, we believe it is crucial that these plants be protected from bird damage. Our primary approach to bird damage management in 1991, will be to train birds to avoid the test plots. We will continue to look for new ways to deter the birds that will minimize the need for trapping. The following approaches will be used during the coming season:

1. Attempts will be made to reduce nesting sites on the campus.

2. Monofilament fishing line and typewriter or computer tapes will be strung around some plots to serve as deterrents at peak times when the grain is most vulnerable.

3. "Scare-eye" balloons will be used as deterrents.

4. A "scare gun" will be purchased and used at peak feeding periods. A scare gun emits a loud whistling sound with streamers that serve as a deterrent.

5. Establishment of nesting sites for sparrow hawks (kestrils) has been underway and will continue to be encouraged as a natural deterrent for pest birds in plot areas.

6. The use of taste and odor repellents will be explored and implemented if they are effective. A study is currently underway with an undergraduate research opportunity program student in the Department of Fisheries and Wildlife under the supervision of Dr. Jim Kitts. The major objective of the test is to evaluate the application of hot pepper solution and grape juice as feeding deterrents.

7. A licensed falconer has been engaged who uses trained hawks to frighten birds away from the plots.

8. Bird distress calls will be periodically played over loud speakers in research plot vehicles as another deterrent mechanism.

9. If necessary, birds will be trapped at times when the grain is most vulnerable (i.e., milk stage). Songbirds will be released. Trapped pest birds will be euthanatized using carbon dioxide according to methods endorsed by the American Veterinary Medical Association.

10. A grant proposal is being submitted to the Humane Society of America to support research to evaluate bird damage control tactics being attempted.

AN HONEST FACE[1]

Professor Jenkins was pleased. As he looked over the computer print-out, he noticed that Steve Adams had scored a 94 his final exam. Throughout the term he had hoped that Steve would pass the course. With a 59 average going into the final exam, Steve had a good chance of passing the course. Since the midterm exam, Steve had stopped by the professor's office several times to discuss class assignments and career decisions. The professor had grown fond of Steve who he viewed as being likable and trustworthy. He felt a momentary sense of satisfaction believing that he had helped a deserving student master one of the department's most difficult courses. As he looked over the other grades he noticed that the class average was up from previous years as was the number of scores in the 90s. Being curious, he looked for other success stories in the class. When he discovered that other marginal students had done well on the exam his satisfaction turned to suspicion. "Why were there so many high grades on the exam?" he thought. When he noticed that Rick Brown had made a 97 on the exam, his fears grew worse. "Had these students cheated on the exam?"

The Student

Steve Adams was a well-mannered and attractive student. He attended class on a regular basis and readily participated in class discussions. He generally sat in the back of the class with a group of fellow fraternity brothers. Unlike his colleagues who tended toward arrogance and indifference, Steve was personable and showed a certain amount of respect for the academic process. Steve's performance on exams and homework assignments was acceptable but not outstanding. He was the type of student that had to work hard to pass his courses. His advisor described Steve as a *good student with an honest face*.

Steve was in his third year of college and was hoping to pull up his grade point average enough to transfer into the School of Business. He was one of many would-be business majors at the University trying to meet the 3.0 grade-point-average needed to transfer into the Business School. Many of these students languished in other schools or majors and tried to take as many business courses as possible. Unfortunately, upper level business courses were restricted to business majors, and Steve was running out of business course electives. Soon, Steve would have to make it into the Business School or choose another major, an option he had never considered and one that his parents would not support. Steve felt that he was under tremendous pressure to get into the business program.

Bill Jenkins admired Steve but didn't think much of his friends and fellow fraternity brothers. He had always been suspicious of social fraternities and had seen much academic talent wasted by the lifestyle promoted by these groups. Only recently had the University began cracking down on hazing and alcohol use by fraternities, much to Bill's satisfaction. In class, Steve sat next to Rick Brown, a fraternity brother. In the few contacts Bill had with Rick, he felt that Rick was arrogant and had a certain amount of contempt for the academic process. Of course, all instructors have personal feelings or opinions about their students. The challenge of teaching is not to let these feelings influence the evaluation process.

The Professor

Bill Jenkins, a tall lanky man with greying hair, was an untenured assistant professor. He carried a heavy teaching load at a state-supported research university. He was well-respected by his colleagues and had earned a reputation as a good teacher. In his second year at the University, he was named *Outstanding Teacher of the Year* by students in his department. He enjoyed teaching and students. He kept an open-door policy and could be found talking to students throughout the day. Having been an average student in college, he tended to give his students the *benefit of the doubt*. On occasions, he has been criticized for pitching his classes toward the average student. He was known to treat his students with respect and trust them to a fault. He felt that if you treat your students with respect, they would treat you with respect. This carried over to his policy on cheating. Not expecting anyone to cheat in his classes, he had no written policy on cheating in his course syllabi.

Bill Jenkins had been teaching for about 4 years and had taught a variety of undergraduate classes. Being rather soft-spoken, he preferred to teach smaller classes (20-40 students). On occasions he would teach classes of 70 or more, which he found to be overly stressful and impersonal. Yet, over the years he had developed some effective large-class techniques. His biggest disappointment about large classes is that the instructor can't get to know his students.

Bill Jenkins' easy-going nature did not translate into easy-going instruction. He had a reputation for being a tough but fair instructor. While giving students the benefit of the doubt, he expected them to work hard. Of course, he was willing to help his students more than many of his colleagues would. For this, some of his younger colleagues thought he was too easy in the classroom. Bill was sensitive to peer comments about his teaching and took steps accordingly. Sometimes, he found himself making his courses more difficult than he felt they should be.

The Course

AAE 438 was an intermediate microeconomics course, taught in the Department of Agricultural Economics. The course had a reputation for being difficult and majors in the department thought of the course as the "rites of passage". The course had two calculus courses as prerequisites. Students who passed the course were relieved but often felt they hadn't learned anything practical. The course covered agricultural economics theory and was designed primarily for majors in the College of Agriculture.

The course had its near equivalent in the Business School, designated as ECN 403. Two similar courses being offered on campus resulted from the rapid increase in the demand for business courses, a shortage of teachers in the Business School, and University efforts to limit enrollments in business courses. Despite some "bad blood" between departments, the Business School would informally recognize course equivalents taught in other departments. Steve and other would-be business majors, begrudgingly took these non-business courses while waiting to get into the business program. Wanting to maintain good relations with the School of Business, the College of Agriculture welcomed these students into their courses. On occasions, would-be business majors were allowed to enroll without the necessary prerequisites.

The Class

This particular class was taught spring quarter and met daily for ten weeks (late March through early June). Fortunately the class was taught at prime time (10:00 a.m.). However, students and faculty alike, often felt rushed and overwhelmed by the pace of the course. Classes became especially tedious near the end of the quarter as the warm weather approached.

The class was taught in a lecture-type classroom with seven rows of fourteen desks each. From day-to-day the number of rows varied from six to eight, depending upon the activities of the previous class or the previous night's cleaning. At full attendance the classroom was crowded. There was little walking room between rows which were often strewn with backpacks. The classroom lacked a raised lectern. Students had trouble seeing the black-board. The instructor had to make a special effort to see students in the back of the room. The room had plenty of full length windows. Sitting in the back of the class was similar to sitting on the back of a *crowded bus*. Giving exams in this classroom setting was a logistical nightmare.

This particular class of 80 students was judged to be excessive for the room size. Students from the College of Agriculture made up about three-fourths of its enrollment. The remaining enrollment was made up of would-be business majors. The class format consisted of daily lectures and objective-type exams. Eighty percent of the grade was based on four one-hour exams, with the fourth exam being given during the final exam period. The remaining grade was based on homework and lab assignments. Lectures were presented by Bill Jenkins. Graduate teaching assistants were employed to grade homework assignments and to help proctor exams.

One of the great ironies of teaching the same course year after year is in being able to predict how students will perform on exams. Bill's experience in teaching AAE 438 was no exception. The average grade on the first exam is around 75. Grades fall on the second exam to around 60 and increase slightly on the third exam to around 65. Average grades rebound on the final exam to around 75. However, some students live dangerously by doing poorly throughout the term in hopes of redeeming themselves at the last hour. The irony of this strategy (from Bill's experience) is that students often misjudge their abilities or the difficulties of learning the cumulative content of a course in a few days. This particular quarter, Bill had the usual number of gamblers in the course.

The Exam

Bill Jenkins did not enjoy giving exams but felt they were a necessary evil. He readily admitted that objective tests are not meant to be a learning experience but are more for the benefit of assigning grades. Thus, he tried to make the test experience as comfortable as possible. Students were allowed to bring food and drink into the classroom, to ask individual questions, and to go to the bathroom on a limited basis. On occasions, Bill would leave the room momentarily to get a drink of water. Students were also given as much time as they needed on the exam. On exam days, Bill Jenkins would ask his teaching assistant to help proctor the exam. Having a second proctor in the room was helpful when Bill's attention was occupied answering student questions.

The final exam was scheduled from 3:00 to 6:00 p.m. on the last day of final exams. With the availability of vacant classrooms, Bill decided to split the class and have the exam given in two separate rooms. He would proctor one room; his teaching assistant would proctor the other. Room assignments were made by splitting the class down the middle; with rows 1-4 meeting in this classroom and rows 5-8 meeting in the seminar room on the next floor. Bill elected to proctor the group in the seminar room, leaving his teaching assistant in the regular classroom.

Thus, the exam was given. As students left the room they commented that the exam was tougher than expected. Some students asked if the grades would be curved and when grades would be posted. He remembered Steve turning-in his final exam and asking how he had done on the exam. Bill glanced at a few key questions on Steve's exam and replied, "Looks like you did a pretty good job!" With a sense of relief, Steve mentioned how much he had enjoyed the course, said good-bye and left the room. As Bill collected the remaining exams, he felt a sense of relief and satisfaction that this had been a good quarter.

Bill followed a set routine for recording grades. Throughout the quarter, he assigned grades by student identification numbers rather than by names. He believed that instructors could be more objective when grading exams if they didn't know whose paper was being graded. Also, when using ID numbers, students tended to visit his office for genuine help, rather than to put in *face time* with the instructor. Unless he made an effort to associate names with identification numbers, he would not know how any one particular student was doing in his class. Often, he was surprised at how well or poorly some students were doing, and how classroom participation was a poor indicator of student performance on exams.

When the final exam scores were returned from the Testing Center, he added them to his computerized spreadsheet. Next, he computed the final numerical scores and arranged them in descending order by student identification numbers. He used this listing to make his cuts between letter grades. Given that Bill's numerical scores were generally lower than average, his cutoff points were typically 5 to 10 points below the University's scale.[2] After making the cuts between letter grades, Bill performed a "reality check" by adding names to the respective scores and grades. Here he looked for consistency within grades and for other information that may affect the cut-off between grades.

During this process Bill discovered the irregularities in scores. There was an unusually large number of high grades on the final exam, especially by some students who had done so poorly throughout the quarter. Not having witnessed any suspicious behavior his first thoughts were that these students had "buckled down" and saved themselves on the final exam. By this time, the evening was late. He entered the final grades on the University's computerized scan sheet and drove by the drop box. Not finding a parking place, he decided to go home and deposit the grades the next morning. That night as he thought about the events of the day, his suspicion turned to doubt and then to panic. "What if there had been cheating on the exam?" he thought. As he became more paranoid he thought "How many students were involved, and how did they do it? Worse yet, could they have been cheating throughout the entire quarter? Why would these students cheat in my class? I trusted them and was more than willing to help them get through the course." As he dwelled on the possibilities, a sinking feeling came over him. "What am I supposed to do now?"

---

[1]*The case was written by Josef M. Broder, Department of Agricultural and Applied Economics, University of Georgia. This case is based on an actual situation that occurred at the author's University. Individual names have been changed to protect the anonymity of those involved. The author is indebted to Steve Turner, Fred White and Joanne Norris for their contributions to the case.*

[2]*The University's guidelines for assigning grades are as follows: scores of 90-100 = A; 80-89 = B; 70-79 = C; 60-69 = D; and 59 and under = F.*

College of Agriculture
Program for Decision Cases
University of Minnesota

Tom and Joan Karen

developed by
Kent Crookston & Mary Hanks [1]

"Probably the worst day for me was when the Sheriff came out in the spring of 1984 to serve papers and we weren't home," recounted Tom as he leaned back in his captain's chair with his hands behind his head. "Sally got home from school just in time to meet him. She was so embarrassed; all the kids on the bus saw it."

"We first started worrying back in '76," he continued. "We didn't get into real serious cash flow problems until '78, '79. We held on into the '80's, but then interest rates climbed."

"Bankruptcy was <u>not</u> an option," Joan interrupted. "We would have taken third jobs. Our neighbors kept telling us to file, even the banker."

Tom Karen and his wife Joan were sitting at their kitchen table talking with a visitor and considering their future. The Karens lived on a farm north of Starbuck (Pope County), Minnesota. They were both 53 years old. Their son-in-law, Brent, who was 28 years old, had just been talking with them about farming the property that Tom and Joan owned. Brent was married to their daughter Sally. Brent and Sally had two children, Brent was employed as an assistant manager in the local grocery store. Brent loved farming however, and on a regular basis had attempted to negotiate some arrangement with Tom and Joan to manage or purchase their property.

Tom and Joan were quite uncertain about Brent's proposal. They too loved the farm, and couldn't imagine living anywhere but in the country. As they sat at the table, the wind whipped swirls of snow across the fields that stretched to the east of their kitchen window. Joan had seen a fox trotting down toward the highway that morning, and while they were still in bed they had heard the clamor of Canada geese in the darkness. Soon the snow would be gone, and they would both be yearning to be out on the land, stirring it up and nursing it into productivity.

---

[1] © 1993, College of Agriculture Program for Decision Cases, 411 Borlaug Hall, University of Minnesota., St. Paul, MN 55108 [phone, (612) 624-1211; E-mail, knut0013@gold.tc.umn.edu]. Kent Crookston, Professor and Head, Department of Agronomy & Plant Genetics; Mary Hanks, Ph.D. and Supervisor, Energy and Sustainable Agriculture Program, Minnesota Department of Agriculture.

But Tom and Joan's yearnings were to go unfulfilled that spring. The fact was that their farm would be left completely untended in 1993. Tom would be "on the road" by planting time, and Joan was scheduled for 35 hours of work per week at the Minneswashka Nursing Home. Since 1986 the entire Karen farm had been standing idle, all of it enrolled in the U.S. Agriculture Department's Conservation Reserve Program known as "the CRP."

Tom and Joan were native to the area and had been high school sweethearts, raised on neighboring farms. "My grandmother actually raised us kids," explained Joan. "With help from my uncle, grandma managed to keep the farm going. It was tough. I pitched silage and milked two dozen cows every day before school."

"I used to go over and help her milk on the weekends," explained Tom. "She had stronger hands and could milk twice as fast as I could."

"Both of our families talked us out of farming," Joan continued. "It was really a hard life back then. I went to school and learned nursing; Tom got into sales. We did pretty good, but both of us knew that we wanted to farm."

"It was 24 years ago, 1969, that we bought our first property," Tom continued. "One hundred and sixty acres. Those were good years. For a while I tried to both sell and farm, I was good at selling, should have stayed with it. But by '76 we had 1700 acres and rented even more. We raised corn, wheat, barley, oats, soybeans and sunflowers. We used the government programs. I quit selling in 1973 and farmed full time. Joan would drive home from work, change her clothes and meet me in the fields with supper. We'd work together 'till past midnight sometimes."

"I kept thinking we were really okay," explained Tom. "But interest rates and land values were doing us in." In 1980 our land was valued at $700 to $1100 per acre but interest rates were over 20%. By 1985 interest rates were back down to 15% but our land values were only $450 to $700 per acre. I owed $275 per acre in 1985; my total interest for 6 months was $42,000. To get a new loan or to refinance I needed 50% equity. That was 50% according to the bank's appraisal which was very conservative. I tried to get some time to meet payments, but everybody was really nervous."

"Hard work and paying your debts was all we ever knew," Joan continued. "Grandma would never have let us even think of living on welfare."

"In '85, with bankruptcy on our doorstep, we decided to try for CRP," Tom continued as he stood up and looked out the window, running his fingers through his graying hair. "We bid $50 per tillable acre (ASC measure). We were accepted for that year and for 9 more years which will carry us through the crop year of 1995. We sold off all our machinery and implements and paid up the bank. I bought my first truck as soon as we were accepted."

"And he's away from home for 335 days a year," complained Joan. "It's *no* fun with him trucking. Certainly not like farming."

"Oh no, not all 1700 is enrolled," Tom explained in response to the visitor's question. "We sold 1200 acres, only 500 are left. We farm 20 acres, have 100 non-tillable, mostly around the house here; we placed 370 in the CRP. "Honey where are those figures we worked up with Brent?" Tom asked as he looked toward the telephone where a stack of papers had accumulated. "My son-in-law thinks he wants to take over the farm," Tom explained. Joan retrieved a sheet of notepad paper and placed it on the table (exhibit 1).

"This could be a real head ache for all of us," Joan offered. "What if our own son-in-law weren't a good manager and we lost our retirement?"

"We wonder now if we'll ever farm again," said Tom wistfully. "Only two years left on our contract, but we can't really make any decision 'til we see the new Farm Bill."

"If they *don't* extend CRP?" Tom repeated the visitor's question. "Sell, invest the money, and sit on it. That's what the economist would say. But where should we put our money? When interest rates are low, you *buy* land. Without CRP, land value and rentals will dip because of surpluses. Maybe farm again. Better to farm and sleep in your own bed; trucking is no way to live. Maybe sell some, and farm some; farming is a good life. Two and a half more years of trucking and the trucks are paid for. I could get back into sales."

"Working where I do, I keep wondering if our health will hold" Joan interrupted.

"If they *do* extend?" Tom's expression quickly changed to one of uncertainty and speculation. "Again we might sell, invest the money and sit on it. It depends on what they tie to the CRP and at what price. Will they guarantee to not renege depending on the whims of congress? Will they allow any flexibility for the farmer in case taxes go up?

"That's a *big* worry," explained Joan. "Some of our land is up in Grant County. They just built a new school up there. It's one of those 'corn field schools[2] ' like the one down by Madison. Before long the State will probably mandate a big school here in Pope County too. Our taxes up in Grant County have doubled in the past 3 years, mostly to pay for that school."

"And now they say the school is polluting Minnewashka Swamp," Tom grumbled. Somebody's going to have to pay for a clean up. Local costs always go up as the State gets in trouble. Too many things can change. The land owner can't afford

---

2 "Corn field school" was what the local people called consolidated schools that were being built on sites that were central to several communities, but within none of them. Students were bussed to these rather large schools from all the communities that were within their radius.

to be locked in for 10 years at a fixed price. If interest rates went back up taxes could kill you. If the cost of living goes up the CRP contract value should go up also"

"What do we hope?" Tom looked at Joan. They both raised their eyebrows and looked more worried than optimistic. "We hope that they extend the program and be generous," answered Tom. "We'd like to be able to put some into grass/alfalfa and manage it for hay. They should not include flat lands. They paid way too much for some lands; 80% of rent should be maximum."

"The public feels that CRP should be public land and open to hunting, etc., for everybody," Tom continued. "It would be cheaper for the government to just buy the land. But how could the government manage all those small pieces? They would have to control the obnoxious [sic] weeds. The government owns too much land in the U.S. anyway."

"Then the land tax base would drop and all other property owners would pay even more tax," Joan warned waving her finger in the air.

"The CRP has been a good program, a smart program," Tom volunteered. "It placed a floor on value, which could have dropped another 20%. That saved some older and poorer farmers from losing their land. It saved tax payer dollars on storage and give-away programs for the commodities that would have been grown on these acres. It saved soil from erosion, and saved land for the day we need it to feed ourselves. This was one of our biggest assets in all wars we have fought and won.

"For sure it was *not* good for some local businesses," Joan pointed out.

"It helped bankers, like ours, get solvent," countered Tom.

"We wonder if there could be some new program that would place young farmers like Brent and Sally back on the land," Joan mused. "Farming is such a good way to raise a family, but it's no good at all if you work for nothing, like surfs."

"We're eventually going to have to see an end to our government's cheap food policy," Tom predicted. "Our government should protect our jobs and products like other countries do. Our productivity saved this country and the rest of the world several times. Cheap commodities shipped in as our jobs are shipped out will some day destroy this country."

"What would you do if you were us?" Joan asked the visitor.

Exhibit 1

[contents of notes and figures worked up by Tom and Brent]

Ten-year goal is to pay all taxes due, and interest. All other money to go to reduce debt on land. C.R.P. Contract is annual $50. x 10 year = $500 per acre.

1986	*Interest on land based at 50% equity was 12.5%*
1992	*Interest on land based at 50% equity was 8.25%*

1986	*Debt was $275 per acre*
1992	*Debt was $242 per acre*

$$loan\ per\ acre\ (\$242\ at\ 8.25\%) = \$19.96$$
$$per\ acre\ real\ estate\ tax = \$9.04$$
$$weed\ control\ per\ acre = \underline{\$1.00}$$
$$Total\ expenses = \$30.00$$

leaving $20.00 per acre for debt reduction, or 4% return on $500 per acre or 7% return on equity of $258 per acre.

1992   Local Bank C.D.s are from 3% to 5.5%.
Land values are slowly rising because CRP has held land off Market.

1985   Rents were $40 - $90 per acre
1993   Rent is $50 - $115 per acre
   $50 is floor placed by C.R.P.

   $115 for irrigated. CRP makes no difference; some land irrigated but has only $50.00 bid.

Taxes
   1986   Real estate tax was $3.66 per acre
   1992   Real estate tax in 1993 is $9.04 per acre due to new corn field school.
   1995 real estate tax will be up [?] $_____
   1995 land value will be? $_____

Estimate $300,000 plus dollars to upgrade and buy machines to farm this land

Betsy's Tempeh

*By Oran B. Hesterman and Gerald Schwab,
Michigan State University*

Gunter knew that he and his wife Betsy had to change direction with their business, "Betsy's Tempeh", but he was still unsure just what the new direction should or would be. The driving force behind the start of the business in the mid-1980's was Gunter's need for a job. "I was 50 years old, had lost my job at Michigan State University (MSU), and it seemed impossible to find another job. My need for income could have been solved different ways, but making tempeh was meaningful work that provided people with healthy food while supporting production of soybeans in a more earth-friendly way." Gunter's early career was in the film-making industry first in Germany and then in the U.S. He came to MSU as a film-related technician in the 1970's, but the source of funds for his position had "dried up" by the early '80's.

The History of Betsy's Tempeh

A few years earlier, Gunter's wife Betsy had found a "how to make tempeh" article in *Organic Gardening* magazine (Exhibit 1) and that was the start of the Betsy's Tempeh story. "Tempeh," Gunter explained, "is a cultured soy product that can be used as a meat alternative in a variety of ways." Gunter is such a "tempeh promoter" that he had authored articles in local Lansing area newspapers trying to educate the public about this "alternative" food (Exhibit 2).

Back in 1978, while still employed by MSU, Gunter had tried his hand at making tempeh, following the directions in *Organic Gardening* magazine. He liked what he made and pretty soon was producing tempeh in six-pound batches in an old refrigerator and sharing it, along with recipes that he and Betsy were developing, with friends. Betsy and Gunter were living at that time on the same 32-acre farm in Williamston, Michigan, on which they live today. Their property is still zoned agricultural and they are operating the business as an

agricultural firm with no need thus far for a special-use permit. Betsy was still working as a librarian at MSU. She believes wholeheartedly in their tempeh business and enjoys showing people how eating healthier can make a difference in their lives.

"The farm was too small to think about making a living with conventional agricultural production, but tempeh seemed to be an opportunity to build a cottage industry on the farm so I could keep busy doing something meaningful," Gunter reminisced. "While experimenting with recipes, we spent three years constructing our food processing building so that we could produce tempeh commercially. We officially opened for business on the summer solstice in 1987" (Exhibit 3).

The rest, as they say, is history (Exhibit 4). For the first two years there wasn't much market growth and Gunter considered shutting the business down in 1989. At about that time, Jane Bush and Gunter discovered each other. Jane owns an apple orchard near Lansing. She distributes apple products and a few other organic products from her farm, as well as Amish chickens and eggs. Jane suggested that she could market Betsy's Tempeh along with her organic apples and apple juice in the Ann Arbor area. With Jane providing marketing and distribution support and Betsy and Gunter busy with in-store demonstrations and samples, sales more than doubled in one year (Exhibit 5). Sales took another leap in 1992 as Jane widened her distribution area and more people found out about Betsy's Tempeh, primarily through word of mouth.

Sales volume remained constant from 1992-93 because Gunter "refused to take on any new customers." The production system he was using had worked well for producing small batches, but was too inefficient to think about producing at a larger scale.

In late 1991, Gunter decided to redesign the tempeh-making process with the goal of producing tempeh on a large scale. "To get ahead as a small business, you need to innovate." And innovate Gunter did. As Gunter redesigned Betsy's Tempeh making process, he procured one patent for the apparatus (Exhibit 6) and another for the method of "a culturing process for food". Gunter claims that Betsy's Tempeh uses a process much different than any other commercial operation and that this is what makes their products better. Gunter and Betsy are convinced that the $15,000 that they spent on the patents was well worth it and will pay off handsomely some day.

Current (1994) tempeh production stood at 6,000-7,000 pounds per year, and the maximum production for their current plant was 150 pounds per day. The total cost of raw products that went into one pound of tempeh was $.25. This did not include overhead such as utilities, land payment, insurance, etc. They could sell the tempeh wholesale to Jane for $2.50 per pound. Jane marked the tempeh up by $.40 per pound, and the retail cost to the consumer was between $4.00 and $4.50 per pound.

Although Betsy's Tempeh was categorized as a small business by anyone's standards, the business was still viewed as quite successful by the local press (Exhibit 7). Success hadn't yet translated into large profits. Betsy and Gunter were still dependent on Betsy's income as a librarian for their living expenses. Unless the business sold soon, Betsy was planning to retire from her library job in about 1½ years. Her real interest has always been in working on the public outreach arm of the tempeh business.

The Dilemma

As the phone rang, Gunter raised his hands in frustration, "If that call is from MSU or someone else who wants to order more tempeh, it will be a disaster! This is a nice test kitchen and we've proven our point, but this facility is too small to meet even our current market demand." As Gunter pondered whether expanding the facility was a good option, he realized that any expansion meant new permits from the Department of Natural Resources to be able to discharge more waste water from the soybean cooking process. Gunter was not at all confident that either the DNR or local planning officials would allow expansion of facilities at their present site, and felt that an industrial site would be needed. The major problems associated with expansion at the present location were increased truck traffic and waste water. Currently, their operation produced 50 gal. of waste water per day. This was not too much of an environmental

load and they were confident that the amount of waste water produced even at maximum capacity of their present facility would not pose an environmental problem or a challenge from the DNR.

However, any major expansion of the business meant not only acquiring a new site but also acquiring new partners and capital. As both Betsy and Gunter considered this, Gunter asked himself and his wife, "given my 61 years of age, is it reasonable for us to take the major next step ourselves or have we done our part and is it time now for some younger folks to take over? If we were still 40 years old, we would love to expand with this product world-wide. We're still great at market demos and want to continue, but I just don't think we have the energy to do the necessary production expansion."

Back in August, 1993, Gunter and Betsy contracted with a broker for one year, and that year was almost up. The broker was attempting to find a buyer for the company. Although several health-food oriented companies liked the Betsy's Tempeh product, none had so far expressed any serious interest in purchasing the company. Gunter realized that the value of the company was not in the current customer list or market demand; the true value, he believed, was in future potential. Their broker had estimated the business sales potential at $50 million per year based on current competition and potential of restaurant and institutional markets. This is why he felt completely justified in the asking price of $1.5 million that the broker had suggested for the two patents, the know how, the trade name, and his consulting time and expertise.

"Should we renew the broker's contract for another year?" Gunter wondered. Gunter and Betsy had just about decided not to seek out new partners and expand the business. "We've burned the candle at both ends for the past 10 years and we're just not willing to do that again over the next 10 years," Betsy stated. But, Gunter added, "if the business doesn't sell, we'll probably continue to make tempeh at the capacity of our present plant just to pay the bills, and that puts us right back in the middle of our dilemma of what to do when a new customer calls."

As Betsy and Gunter looked at the telephone, they both hoped that it wouldn't ring. Unless, of course, it was the broker with news of a buyer.

Study Questions

1. Who is the decision-maker?

2. What is the dilemma?

3. What are the decision-maker's objectives?

4. How profitable is Betsy's Tempeh? What are your future projections?

5. What options does the decision-maker have?

6. What should he/she do and why?

Exhibits

1. "Calling All Tempeh Lovers," *Organic Gardening and Farming,* June 1977, pages 108-111.

2. "The Temptation of Tempeh," advertisement in *Nexus News.*

3. "It's Not a Burger, It's Betsy's Tempeh—and It's New on Mid-Michigan Market." *Lansing State Journal,* June 29, 1987.

4. "Betsy's Tempeh: A Brief History and Update."

5. Betsy's Tempeh Sales 1988-93.

6. United States Patent No. 5,228,396, "Apparatus for Culturing Plant Materials as Foods." Inventor: Gunter Pfaff. July 20, 1993.

7. "Triumph for Betsy's Tempeh." *Lansing State Journal,* March 10, 1994.

EXHIBIT 1

Organic Living

Calling All Tempeh Lovers

*Here are more ways to get started with this nutrient-rich
soybean food, along with recipes and variations using grains.*

HWA L. WANG, E. W. SWAIN
AND C. W. HESSELTINE

OVER 3,000 PEOPLE have asked us for tempeh starter, the dried culture that gets this soybean fermentation going, since our offer first appeared in the January '77 OGF. We sent everyone a packet with enough to make several batches, but we were just too busy getting the starter in the mail to answer all the questions about making tempeh.

More than anything else people wanted to know if there wasn't a way to keep their tempeh going at home. We didn't have a good answer to that one — ours is always made under laboratory conditions. So we began to work out a way to do it. Now that we are satisfied with a technique, we want to pass it along to everyone. We'll also take this opportunity to answer the other most asked questions.

There are two ways to maintain a tempeh supply. The beneficial mold

that makes tempeh can keep growing by getting more fresh soybeans to work on and by making spores. The only practical problem with simply adding some active tempeh to your next batch of beans is the remote possibility of bacterial contamination. A few bacteria are bound to get in, and their numbers multiply each time a new batch is made. Eventually these outsiders prevail, and the beans won't form into a tempeh cake.

There isn't much you can do to keep them out. But here is the procedure that has given us good results: First prepare a pound of dry soybeans as directed for making tempeh. In a blender, mix an ounce of active fresh tempeh with one to two tablespoons of boiled but cooled water to make a thin paste. Then mix that paste with the prepared beans, and pack and incubate them as usual. Try to reduce to a minimum the time any of these ingredients are exposed to open air. In successive batches, watch the quality of your starter tempeh for bacterial contamination very carefully, or you may waste a pound of beans. A good tempeh cake is

Drs. Hwa L. Wang, E. W. Swain and C. W. Hesseltine have been testing tempeh production and use at the Northern Regional Research Center, Agricultural Research Service, USDA.

Tempeh is a fine snack, replacing sugar-laden cavity-causers with good-tasting protein for a real pick-me-up. Tempeh has the natural, crunchy, nutty flavor that candy bars are always trying to imitate.

This incubator is used in commercial production of tempeh by Gale Randall in Omaha, Nebraska. Randall has a small home business producing tempeh, and custom-built his incubator to meet his large-scale needs.

clean-smelling and rather firm.

Another way, which we much prefer, is to make your own special starter with rice. You need a pressure cooker, mason jars, and milk filters which will replace the lids of the mason jars. If milk filters are not available, they can be made by putting cotton between two layers of gauze. The thickness of the ·cotton is about ⅛ to ¼ inch, and it serves to keep dust and bacteria out of the jar, but allows the mold to breathe. You can then make your own starter by following these simple steps:

1. Place ¼ cup of white rice in a mason jar (pint or quart), add one ounce of water, mix well, cover with milk filter in place of the lid, and put the screw top on.

2. Let the jar stand at room temperature for one hour, and shake it every 10 to 15 minutes so that the rice will take up water uniformly.

3. Cook in a pressure cooker for 20 minutes with 15 pounds of pressure.

4. Shake the jar to break up the rice lumps, cool and add ⅛ teaspoon of tempeh starter. Put back the filter and

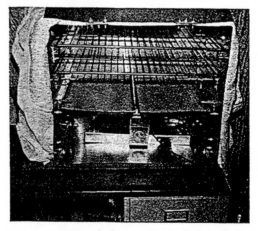

screw top, and then mix well by shaking.

5. Lay the jar on its side to spread out the rice, and place it at 86 to 88 degrees F, for four days. At that time, rice should be covered with black spores. Pulverize the spore-covered rice in a dry, clean blender for one to two minutes until it looks like uniform dark-gray granules. The starter is now ready for use. Put the rest of the starter back in the mason jar, and cover the jar with filter, lid and screw top. Then keep the jar in a freezer; the starter can be kept for many

months — it will not lose vigor.

This starter is much more potent than the one that we sent to you, because we have diluted ours with wheat flour. You really need only ⅛ teaspoon of this starter to make tempeh from one pound of dry soybeans. Mix a very small amount (⅛ teaspoon) of the starter in two teaspoons of cool, boiled water, and add to beans prepared for tempeh. Cover the mixing bowl with a plate or a piece of plastic film, and vigorously shake the bowl for ten to 15 seconds to thoroughly mix the starter and beans. Pack and incubate as usual.

TEMPEH VARIATIONS

Tempeh-type products can be made from wheat, oats, rye, barley and rice, as well as from any of these in combination with soybeans. They all possess a pleasant odor, a desirable color, and a good taste. All grains are slightly cracked or pearled, rinsed with water, soaked for 30 minutes at room temperature, and boiled for ten to 12 minutes. Cool, inoculate, pack and incubate as you do with soybean tempeh. Other beans probably would be good to make tempeh-type products, but we have not tried them.

It is not necessary to scrape any of the mold off — it disappears in cooking. But you must cook tempeh before eating. Tempeh slices can be deep-fat fried, sautéed or roasted. One reader wrote to say that she sautéed tempeh cubes and used them in salads as croutons. Tempeh patties cooked like hamburger have been used in sandwiches. It is also good in your favorite soups or casserole dishes.

Tempeh does not have to be cooked and eaten the day it is made. It can be kept in a refrigerator for a day or so, or in a freezer for a few months. After thawing, tempeh tends to break up when sliced. So, if you are going to freeze tempeh, it should be cut first.

Many asked if they could use the soy grits widely available in stores. Al-most always these are defatted grits and are not satisfactory. The most suitable form of soybeans for making tempeh is the full-fat, cracked bean (about four to five pieces per bean). Smaller grits, about half the size of a rice grain, can also be used to make tempeh; however, a greater amount of protein and other soluble nutrients is lost during cooking.

NUTRITIONAL VALUE OF TEMPEH

The amount of proteins and the protein value of soybean tempeh are the same as that of the soybeans. Vitamins such as niacin and riboflavin in tempeh are greatly increased; thiamin or vitamin B_1 may be slightly decreased. As for vitamin B_{12}, there is not enough information to indicate its presence in tempeh or soybeans. At the present, neither tempeh nor soybeans should be considered as a source of vitamin B_{12} in a vegetarian's diet, but we are looking into the matter.

EXHIBIT 2

News News

The Temptation of Tempeh

by
Gunter Pfaff, Maker of Betsy's Tempeh

TEMPEH (pronounced Tem-pay) is a new word for many people in our culture; even when we serve samples of Betsy's Tempeh in the Food Co-op, as many as 65% of this food conscious group of consumers have never heard of it. Tempeh is an Indonesian word (tempe) and refers to a variety of legumes that have been cooked and bound together with a white mycelium (mushroom culture) during a 24 hr. incubation period.

Today tempeh is mostly made with soybeans as the main ingredient; other grains are often added to achieve a certain flavor and texture.

Nutritionally it is an ideal high protein cholesterol free alternative to meat for any style of cooking you might favor. Most often it is found in the form of 8 oz. "cakes" which can be sliced or cubed and served fried, baked or steamed. Our own version of tempeh is made with a small amount of barley added to the soybeans; it is precooked and ready to use. Betsy's Tempeh comes in either 8 oz. grated form or (12) 2 oz. patties. It is extremely easy to use, for example, just take the 8 oz. grated tempeh from the package and add it to a 32 oz. jar of your favorite spaghetti sauce, heat and serve over noodles (or pizza). By itself tempeh is fairly mild tasting but easily adopts flavors--most often tamari or soy sauce is used to flavor it.

All the tempeh I have seen on the freezer shelves of local stores is made with organically grown soybeans and grains which means that whenever you serve tempeh, you are supporting an ecologically sound use of our land while keeping yourself healthy at the same time.

Currently about 2 million pounds of tempeh are made in this country yearly by about 35 producers; it is mostly sold in Food Co-ops and health food stores. In its country of origin, Indonesia, where it has been a staple in the diet for hundreds of years, it is produced in about 41,000 shops.

There is a good chance that because of its amazing qualities, Tempeh could, within a short time, be commonly available in this country and almost everyone would soon experience *the temptation of tempeh*.
For more recipes:
TEMPEH COOKERY by Colleen Pride: The Book Publishing Co., Summertown, TN 38483
THE BOOK OF TEMPEH. The Delicious, Cholesterol-Free Protein: 130 recipes by William Shurtleff and Akiko Aoyagi: Harper Colophon Books (available from Soyfoods Center; PO Box 234, Lafayette, CA 94549)

EXHIBIT 3

Today   Lansing State Journal ■ Monday, June 29, 1987

It's not burger, it's Betsy's Tempeh — and it's new on mid-Michigan market

No cholesterol, high in protein, say producers

By PAMELA JAHNKE
Lansing State Journal

PERRY — "Betsy's Tempeh" won't appeal to everyone — but then, the locally produced product is not being targeted at just anyone.

"It's basically for people who want to decrease their meat intake or get off meat altogether," said Betsy Shipley, who is marketing the product with her husband, Gunter Pfaff, both of whom are vegetarians.

She said the cultured soybean product will be available this week in patties or a grated form at area food co-ops and food-buying clubs.

Shipley, a library clerk at Michigan State University, believes the product is the only one of its kind in the Lansing area, and possibly in Michigan.

It will be produced at the couple's 32½-acre farm in Perry in a 900-square-foot farm kitchen designed by Pfaff, a retired MSU employee. About 500 pounds of "vegetarian hamburger" will be produced there a month, Shipley said.

Shipley added that tempeh is a good alternative to meat, and provides another option for those who fear they are eating too much cheese. Unlike cheese, the product does not contain any cholesterol or animal fat, Shipley said.

Tempeh, which has been around at least a few hundred years, originated in Indonesia, where it is the main staple of the native diet, Shipley said. It is a fermented food bound together by a dense mold.

Although it has been available locally in 8-ounce squares, Betsy's Tempeh is packaged in a more convenient form for the con-

Lansing State Journal/ROD SANFORD

Betsy Sipley and Gunter Pffaff hold a platter of "Betsy's Tempeh," a food the couple will produce in this production house near their Perry home.

sumer, Shipley sys.

In addition it is made with soybeans and barley — both of which are grown on organic farms in Michigan, Shipley said. The barley gives the tempeh a "nuttier" flavor, she added.

The 2-ounce patties will come 12 to a 1½-pound package and the grated form will come in 8-ounce packages. Both packages will include nutritional information and suggested recipes.

One patty has 110 calories, 12 grams of protein and 4 milligrams of sodium. A 4-ounce serving of the grated form has 220 calories, 24 grams of protein and 8 milli-

grams of sodium.

The wholesale price of either form of tempeh is $2.26 per pound, Shipley said.

The patties can be baked, broiled or fried while the grated form can be used as a ground beef subsitute in dishes like chile, soups or eggrolls.

Tempeh is versatile enough to be flavored with just about anything ranging from curry powder to Mexican spices, Shipley said.

"It basically absorbs whatever you want to give it," Shipley said.

She said tempeh can be stored in the freezer for up to six months and thawed by being placed in a

microwave or a steamer.

For five years, Shipley and her husband had been planning to create another cottage industry on their organic farm. They also grow and sell elephant garlic, a milder-tasting and larger form of garlic.

EXHIBIT 4

Betsy's Tempeh® :

A Brief History and Update.

1980-84	Betsy Shipley and Gunter Pfaff begin to develop an easy to use, ready to eat tempeh made with organically grown soybeans and barley. At this point it is produced and used in the home and tried out on friends. Plans are made to build a small food processing facility on our farm. The goal: to commercially produce the same quality of tempeh as could be made at home.
1984-87	With lots of elbow grease we build a 1000 sq ft building, using a variation of a ferrocement construction technique. (Pressing mortar into wiremesh). This method is ideal since the building needs to be rodent proof on the outside and humidity proof on the inside. (Our inspector loves it !)
1987 June	Our OPEN HOUSE is catered with many memorable tempeh dishes; we begin deliveries to the local Food Coops and restaurants. The product finds a good reception, but our market needs to be larger.
1989	A very small distributor takes our product to the regional market and we do in-store demonstrations and explore working with institutions.
1991	"Word of Mouth" boosts our sales 48% by July. New production equipment is designed and installed by the end of the year.
1992/3	We file our patent for our novel production process; the "apparatus" patent is published in July of 1993. The "method" patent is submitted. We sign up with an experienced broker because we want to sell the patents due to our age. Sales keep increasing, mostly by word of mouth - 68% by the end of the year.
1994	By the end of March we are 62% ahead of 1993 and Gunter is " maxed out " and prays for finding a suitable buyer very soon.

For approx. two decades, the most commonly seen tempeh in Food Coops and Natural Foods Markets has been what is called the 8 oz. "brick". (A solid piece, incubated in the perforated bag in which it is sold, most often un-pasteurized). This type of product could find a ready market only with a small knowledgeable group of customers, it could not reach the much larger group of people who are just as motivated to eat healthier but have less know - how and who are not willing to steam their "brick" for 5 - 10 min. first.

When Betsy's Tempeh was able to offer a ready-to-eat tempeh in a convenient shape, patties and grated, we not only got rave reviews from those already familiar with tempeh, but we ended up attracting large numbers of newcomers to this delicious and useful alternative.

There is no doubt that Betsy's Tempeh is ready to become much more widely known and used. This is true not only for the Food Coops, Natural Foods Markets and certain health oriented supermarket chains, but also for large institutions, mainstream supermarkets and especially restaurants.

Most of the changes occuring in the market place are customer driven --
Now is the time to produce what people want.

Betsy's Tempeh® is ready to become a household word.

"Indonesian Bean Steak"

Betsy's
TEMPEH®
SIMPLY THE BEST
ALTERNATIVE

A High Fiber, High Protein, Cholesterol Free Cultured Soyfood

For Burgers, Kebabs, Pizza, Chili, Soups, Tacos, Fajitas, Spaghetti, Stir Fry, Salads, Sloppy Joes, Stuffed Peppers, Dim Sum, Reubens, Spreads...

Made from organically grown soybeans and barley with a mushroom-like culture

Good For You And The Planet

EASY TO USE - READY TO EAT
PATTIES ✳ GRATED

EXHIBIT 5

Betsy's Tempeh Sales 88/93

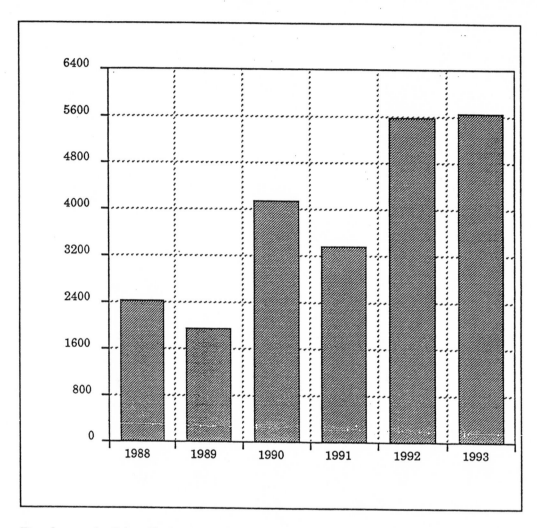

By the end of April 1994 production ran 35% ahead of 1993.

EXHIBIT 6

US005228396A

United States Patent [19]

Pfaff

[11] Patent Number: 5,228,396

[45] Date of Patent: Jul. 20, 1993

[54] **APPARATUS FOR CULTURING PLANT MATERIALS AS FOODS**

[76] Inventor: **Gunter Pfaff**, 14780 Beardslee Rd., Perry, Mich. 48872

[21] Appl. No.: **871,065**

[22] Filed: **Apr. 20, 1992**

[51] Int. Cl.⁵ A23L 1/20; A23L 3/00; A23L 3/10

[52] U.S. Cl. 99/470; 99/453; 99/467; 99/473; 99/483; 99/536; 261/121.1; 261/124; 422/300; 422/305; 422/307

[58] Field of Search .... 99/323.1, 453, 467–470, 99/473, 476, 483, 484, 487, 516, 534, 536; 366/101, 106, 107; 261/124, 126, 121.1, DIG. 7, DIG. 16, DIG. 30, DIG. 76; 219/401; 422/297, 300. 302, 305, 307, 231

[56] **References Cited**

U.S. PATENT DOCUMENTS

1.513.174	10/1924	Kruger .	
3,228.773	1/1966	Hesseltine et al. .	
3,243.301	3/1966	Hesseltine et al. .	
3,874,279	4/1975	Sakita et al.	99/516 X
3,933,953	1/1976	Leva	261/113
3,981,234	9/1976	Nelson et al. .	
4.013,869	3/1977	Orts .	
4.076.617	2/1978	Bybel et al.	261/124
4,189,504	2/1980	Jimenez	99/536
4,248,141	2/1981	Miller, Jr. .	
4,534,283	8/1985	Nakamuta	99/483 X
4,563,277	1/1986	Tharp	210/124 X
4,769,221	9/1988	Marihart	422/231
4,771,681	9/1988	Nagata	99/453 X
4,848,216	7/1989	Robau .	
5,015,394	5/1991	McEllhenney et al.	261/124 X
5,133,249	7/1992	Zittel	366/107 X
5,142,969	9/1992	Chun	99/470

FOREIGN PATENT DOCUMENTS

1274428	8/1968	Fed. Rep. of Germany	99/483
3604808	8/1987	Fed. Rep. of Germany	99/470
2145004	3/1985	United Kingdom	261/124

OTHER PUBLICATIONS

Discusses inoculum enrichment by natural selection as an alternative to pure culture starters (Unknown as to Source and Date).

Shurtleff & Aoyagi (1980) Tempeh Production, The Book of Tempeh: vol. II, published by New–Age Foods pp. 46–50.

Soyfoods, Vegetarian Times, pp. 35–39, Nov. 1987.

Primary Examiner—Timothy F. Simone
Attorney, Agent, or Firm—Ian C. McLeod

[57] **ABSTRACT**

An incubator apparatus (10) for preparing an aerobically cultured plant material, such as a soyfood substrate (11), inoculated with a beneficial microorganism to form a cultured food, such as Tempeh, is described. The incubator apparatus is comprised of a water tray (13), which provides a water bath (15) for heating the inoculated soyfood substrate loaded in shallow metal trays (45), preferably stainless steel trays. The trays are then supported on tray racks (43) that are mounted in the water bath so that the trays are partially immersed in the water bath. The trays holding the inoculated soyfood substrate are then sealed in the water bath by a cover (61). The cover mounts over the tray racks and is partially immersed in the water bath to seal the cover over and around the trays filled with the inoculated soyfood substrate. An aerating supply system (31) provides for the aerobic culturing of the soyfood substrate while a circulating pump (100) provides a uniform distribution of the water bath throughout the water tray. A sensor (132) and a controller (104) actuate a temperature control system to effect heating and cooling of the water bath as needed to promote the growth of the microorganisms on the soyfood substrate. Later, a heating system (17) is used to elevate the temperature of the water bath to stop the culturing process by killing the microorganisms and to pasteurize the cultured food. The cultured food is then cooled and removed from the trays as a ready to eat food that can serve as an alternative to meat. The Tempeh is high in protein content and high in fiber without having cholesterol.

22 Claims, 6 Drawing Sheets

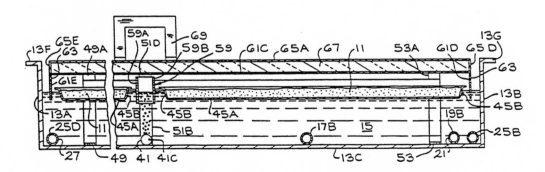

EXHIBIT 7

Triumph for Betsy's Tempeh

Lansing State Journal
March 10, 1994

Perry-based operation builds soyfood sales

By GISGIE GENDREAU
Lansing State Journal

PERRY — It all started 18 years ago when Gunter Pfaff received a tempeh-making kit for his birthday.

Now, Pfaff and Betsy Shipley are having a hard time keeping up with demand for their square soy patties. They're looking for a buyer for Betsy's Tempeh burgers so they can be produced internationally.

Tempeh, pronounced "tem-pay," is a cultured soyfood that originated in Indonesia. A mushroom culture makes the soy into tempeh.

SMALL BUSINESS

When Betsy's started in 1987, the couple had a hard time selling their soyfood patties in what Shipley calls "meat and potatoes country."

Since then, the Perry-based operation's sales have grown significantly, including a 68 percent jump from 1991 to 1992. That's from 4,760 to 7,000 pounds of tempeh in a year, all made by Pfaff.

Their customers increased again in September when Michigan State University and Whole Foods Store in Ann Arbor filed orders.

"We're maxed out," Pfaff said. "People always dream about success, and now it's close to a nightmare."

Pfaff and Shipley created and patented a procedure to produce the tempeh in a structure near their home. A sign reading "Only the impossible is worth doing" is on its door.

"I needed a job; people eat every day," is how Pfaff describes Betsy's beginnings. He believes the product's success is partly due to a trend toward healthier eating.

"It's changing fast. There's enough information coming out in the media that people should be eating healthy."

The 132-calorie patties have 14.4 grams of protein, 9.6 grams of carbohydrates and no cholesterol. They're sold in packages of 10 or shredded for dishes such as tacos and spaghetti sauce, Shipley said.

Betsy's tempeh burgers are popular at the Hearthstone Vegetarian Restaurant in East Lansing, owner Bruce Roth said. "They've been received well. We're selling lots and lots of them."

The restaurant, 208 M.A.C. Ave., also serves the burger in chili, as sloppy joes and as sausage in pizza. The burgers, served with chips, cost $2.50, Roth said.

Roth added the product to his Lansing restaurant, the Hearthstone Bakery, about four years ago.

It has been popular because it is high in protein and takes on the flavor of whatever is added to it, he said. Unlike hamburger, it isn't greasy, Roth said.

"You don't get it evaporating away in grease," he said. "It works out real good for us."

It's also working out for Michigan State University's residence hall cafeterias. There, patties are cubed and used in a vegetable stir-fry up to six times each semester. Food Services coordinator Michael Rice said.

Demand for vegetarian dishes is growing, and products such as the soy patties help the university keep up, he said. From 10 to 12 percent of the 1,400 students served each day pick a vegetarian dish. "Every year, they're more and more popular," Rice said.

Some vegetarians specifically ask for tempeh dishes at the Travelers Club International Restaurant & Tuba Museum in Okemos, said D. Thomas, a manager and chef. The club has been using Betsy's product for at least four years, he said.

"It's a perfect meat substitute," Thomas said. "That's our protein."

The club serves tempeh with eggplant, tofu and in oriental dishes, he said. "We even serve it up in a sloppy joe," Thomas said. "It doesn't taste like a meat sloppy joe, but it's a nice and spicy little burger."

Lansing State Journal/CHRIS HOLMES

Gunter Pfaff and Betsy Shipley make tempeh at their home near Perry. They hope to sell patents assigned to the equipment and recipe they've developed.

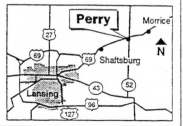

Betsy's Tempeh

Tempeh, an Indonesian soyfood, can be used as a meat substitute. To get a free recipe book, send a self-addressed, stamped envelope to:
Betsy's Tempeh
S&P Farm
14780 Beardslee Road
Perry 48872

Lansing State Journal

Minto-Brown Island Park: Farming the Urban-Agricultural Interface

*By D.L. Taack, H. Murray and S.R. Simmons,
University of Minnesota*

One observer compared the situation at Minto-Brown Island Park to a "powder keg near a lit match." Another stated, "It was like the spotted owl, only on a smaller scale." (Savonen, 1993). Minto-Brown Island Park was unusual as agricultural fields were incorporated within the park grounds. Visitors to the park in Salem, Oregon had observed a substance being sprayed on a waterway near agricultural land located in the park. Assuming the park's tenant farmer was irresponsibly handling agricultural pesticides, the visitors reported that information to their local Audubon Society chapter. In 1989 the Salem, Oregon chapter of the Audubon Society published a report (Doherty, 1989) criticizing the use of pesticides on farm acreage within the Minto-Brown Island Park of Salem. The report recommended that the Salem Parks Department and the city of Salem ban the use of pesticides suspected of causing long term health or environmental damage. R.G. Andersen-Wyckoff, then president of the Salem Parks Advisory Board, felt the situation would develop into a major confrontation between agricultural and environmental interests unless it was quickly resolved. Another member of the parks board, who was also a member of the local Audubon chapter, had pressed Andersen-Wyckoff insisting that he implement the report's recommendation to ban pesticide use on the park farm.

Salem is the capital of Oregon, the county seat of Marion County and part of one of the most productive and diverse agricultural regions in the state, the Willamette Valley. Marion ranks as Oregon's number one agricultural county based on farmgate sales. Andersen-Wyckoff knew that decisions regarding pesticide use in the park could have wide-ranging implications for agriculture and pesticide use policies throughout the state.

In 1989, Andersen-Wyckoff discussed the situation with Rob Miller, a local farmer who had been active in agricultural issues in the area for some time. Mr. Miller provided Andersen-Wyckoff with a list of people who he knew had an interest in the outcome of the pesticide decision. This list included farmers, park-users, environmental groups including the Audubon Society, local businesses, and land owners in the area. These groups represented a variety of interests. For example, the Oregon Tilth organization worked toward organic certification for farmers, while Oregonians for Food and Shelter favored conventional agriculture and opposed further regulation of agricultural chemicals. The Oregon Agricultural Chemical Association was also interested in cases involving public concern and perceptions related to pesticide use. The park's tenant farmer was a member of the Northwest Food Processors Association, a research, marketing, and advocacy organization. Food processors were still reeling from the Alar controversy earlier that year and concern about public perception regarding pesticides was high. Oregon State University Agricultural Experiment Station and Cooperative Extension Service personnel also had an interest in the outcome of the park board's decision as an agricultural issue. This was clearly an issue that touched many elements and interests within the Salem community and beyond. Andersen-Wyckoff knew that great care would be required to resolve this dilemma while maintaining support of the many diverse parties interested in the outcome of the Park Advisory Board's actions.

Background on the Decision Maker

R.G. Andersen-Wyckoff had been a member of the Park Advisory Board for six years and had been president of the Board for two years before the Audubon Society issued its report. In 1989 he was nearing the end of his term as president and was entertaining thoughts of becoming a candidate for mayor in 1990. A local business owner, he had no background in agriculture and felt his views were more those of a "conservationist" rather than a "preservationist." Andersen-Wyckoff felt that the principal components of Minto-Brown Island Park, wildlife conservation, recreation, and agriculture, were "like three legs of a three-legged stool." Each "leg" contributed to and was necessary for the others. He believed that the agricultural component of the park contributed to its uniqueness and helped Salem citizens relate to and identify with the agricultural heritage and economy of the Willamette Valley.

Background on the Park and Farm[1]

In 1857 Isaac "Whiskey" Brown homesteaded an island in the Willamette River near the present community of Salem. In 1861 the worst flood in Salem history changed the course of the Willamette River and caused Brown's property to be joined with another island, and the southwest edge of Salem (Savonen, 1993). Another settler, John Minto, purchased a parcel on the other island, which was now also connected to Brown's. Both the men cleared their lands, converting them into farmland which has remained in agricultural production since that time (Exhibit 1).

In 1970 the City of Salem and Marion County acquired 860 acres of these lands to create a park and wildlife refuge. The park was subsequently expanded to 892 acres. The city named the park after the original settlers, Minto and Brown. Beginning in 1970, 240 acres of the park area were leased to area farmers for agricultural use. Farmer Ken Iverson had been the sole tenant since 1986. Iverson grew commercial potatoes, beans, and sweet corn (Exhibit 2). As part of his lease agreement he planted corn and cover crops for waterfowl. The lease was due to expire September 30, 1991.

The park was also under the jurisdiction of the National Parks Service as it was located in the National Wildlife Refuge of the Willamette Valley. Over 175 bird and 35 mammal species had been identified within Minto-Brown Park. Dusky Canada Geese, a rare species, regularly over-wintered in the park. The geese and waterfowl were a prime attraction of the area, as well as the scenery, bike trails, and other recreational facilities.

[1] Unless noted, factual information included in this section was obtained from Gredler (1990).

Primary recreational users of the park were runners, walkers, bicyclists and bird watchers. On any given day, over 100 runners and walkers were known to use the park facilities.

Agriculture also played a significant role at Minto-Brown, allowing urbanites to better view the cycles of agriculture such as plowing, planting, and harvest. In addition, the farming practices provided excellent food and shelter for some of the wildlife. The park's board did not have the resources to provide the necessary cover crops. Therefore, without the farmer's inputs, much of the land would revert to bramble and blackberries, of little use to the waterfowl.

The Pesticide Review

The 1989 Audubon Society report which proposed banning pesticide use on the Minto-Brown Island Park farm was prepared by Maura Doherty, an industrial hygienist. The report had been jointly funded by the city of Salem and the local chapter of the Audubon Society. The Audubon Society was particularly concerned with the park's use of dinoseb (sold under the brand names "Preemerge 3" and "Dinitro General"), a herbicide that had been under emergency suspension by the Environmental Protection Agency, since October 1986 (Exhibit 3). Dinoseb could be used as a preemergence herbicide and for vine killdown in potatoes. The EPA stated that exposure to dinoseb presented "imminent" and "unacceptable" risks of detrimental health effects to exposed persons. These risks included sterility, birth defects, and other health hazards.

The Audubon Society report (Doherty, 1989) also indicated that a total of 24 pesticides or growth regulators could have been used in the park and evaluated each of them for their possible effects on health and the environment (Exhibit 4). The report concluded that the majority of these chemicals could cause long term health problems in experimental animals. EPA testing had not been completed on eight of the pesticides at the time the report was prepared. Fifteen of the chemicals had been found in contaminated water supplies in other parts of the United States. The Audubon researchers had not spoken to the farmer regarding specific pest management strategies used on the property. The report did not indicate how frequently or widely each chemical had been used by the current or previous farmer(s). Neither did it consider pesticide use on the Salem Golf Club or on the privately-owned Boise Cascade properties adjacent to the park.

All of the pesticides were labeled by the EPA at the time they were used by the farmer. Although dinoseb could no longer be purchased, existing supplies could still be used in accordance with the label. It was under this provision that dinoseb had been used on the Minto-Brown farm since 1986.

The Audubon Society report recommended that the city of Salem take action to ban the use of all pesticides in the park, citing increasing concern among "many park users" (Exhibit 5). The pesticide review was the first phase of a three-phase process planned by the Audubon Society. Phase two was to evaluate alternatives to chemical pesticide use in the park and phase three would evaluate the feasibility of developing a compost program using vegetation from the park grounds.

The Decision

Andersen-Wyckoff had to decide what action to take to alleviate tensions among environmental and agricultural interests in the community and resolve differences regarding use of pesticides on the Minto-Brown Park farmland. He knew that members of the nonfarming public expected farmers to protect the environment and conserve natural resources without raising the cost of food. Some clearly regarded pesticide use on the farm as hazardous to park users and detrimental to wildlife and the environment. They expected Andersen-Wyckoff to act favorably on the Audubon Society report's recommendations. On the other hand, some representing farming interests regarded the report as alarmist and poorly researched and expected Andersen-Wyckoff to protect their right to utilize legal pesticides when the situation merited.

There were many questions in Andersen-Wyckoff's mind as he weighed how he should respond to the Audubon report.

- Were the concerns of the Audubon Society regarding adverse pesticide effects on wildlife valid?

- Would restricting pesticide use on the Minto-Brown farm have symbolic and negative repercussions for other farmers in the region or in other parts of the state?

- Had pesticide use by the tenant farmer been a hazard to park visitors?

- Was there a way to manage the agricultural part of the park in such a way that all of the diverse interests and stakeholders, farmers, consumers, park patrons, and environmentalists would be satisfied?

- Could he turn this situation into a positive one by uniting these diverse interests rather than heightening confrontation?

References

Doherty, M. 1989. "Pesticide Review: Minto-Brown Island Park, Salem, Oregon, 1984-1988." Salem Audubon Society, Salem, OR.

Gredler, G. 1990. Minto-Brown Agricultural Project: Current Assessment. Oregon State Univ., Extension Service, Marion Co., OR.

Savonen, C. 1993. "The urban-agricultural fringe." *Agricultural Engineering*. Mar 1993, pp. 17-20.

Exhibits

1. Map of Minto-Brown Island Park.
2. Crops and crop rotations on the Minto-Brown farmland, 1986-1989 (Gredler, 1990).
3. Fact sheet for dinoseb from the Pesticide Review for Minto-Brown Island Park (Doherty, 1989).
4. Toxicity and water contamination information for pesticides used in Minto-Brown Island Park (Doherty, 1989).
5. Excerpts from the Executive Summary of the Pesticide Review for Minto-Brown Island Park (Doherty, 1989).

EXHIBIT 1

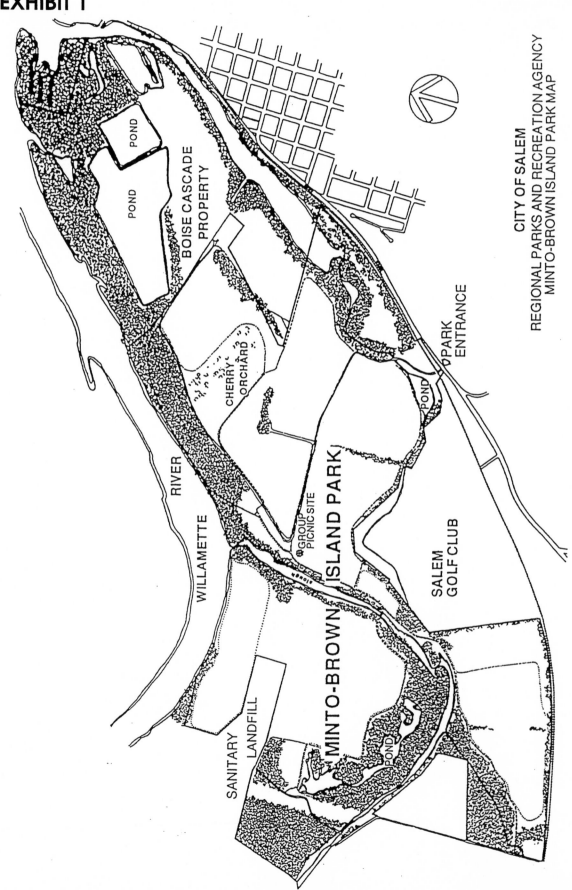

CITY OF SALEM
REGIONAL PARKS AND RECREATION AGENCY
MINTO-BROWN ISLAND PARK MAP

EXHIBIT 2

CROPS AND CROP ROTATIONS ON THE MINTO-BROWN FARMLAND, 1986-1989

(Gredler, 1990).

FIELD 1

1986 potatoes
1987 beans
1988 corn
1989 corn

FIELD 2

1986 potatoes
1987 1/2 fallow 1/2 beans
1988 potatoes
1989 beans

FIELD 3

1986 beans
1987 potatoes
1988 beans
1989 potatoes

FIELD 4

1986 potatoes
1987 potatoes
1988 corn
1989 corn

FIELD 5

1986 potatoes
1987 fallow
1988 corn
1989 corn

FIELD 6

1986 bird feed
1987 potatoes
1988 corn
1989 corn

EXHIBIT 3

FACT SHEET FOR DINOSEB FROM THE PESTICIDE REVIEW FOR MINTO-BROWN

ISLAND PARK (Doherty, 1989)

LEGAL STATUS

In October 1986 EPA announced an emergency suspension of dinoseb use. Prior to this approximately 180 products contained dinoseb and over 11 million pounds were sprayed annually in the U.S.

TOXICITY

Acute

The oral LD 50 of dinoseb was between 20 and 60 mg/kg or highly toxic. 100% of the dose applied to skin could be absorbed.

Chronic

Dinoseb lowered the sperm count in exposed animals and caused teratogenic and reproductive effects. Dinoseb was listed as a potential human carcinogen and caused the development of cataracts at low exposure levels. Dinoseb also damaged the immune system, liver, kidney, and spleen.

POPULATION AT RISK

Dinoseb was highly toxic to fish and wildlife causing fishkills and field kills of pheasants and songbirds. Dinoseb was banned from use because of the potential harm to the reproductive system of workers and exposed members of the public.

ENVIRONMENTAL FATE

EPA noted that this chemical was a Priority I chemical able to leach from soil into water. Dinoseb can persist on crop treated soil from 2 to 4 weeks. Ontario, OR wells show persistent contamination with dinoseb. A British Columbia report noted that dinoseb showed residual action in soil for approximately 1 month or less.

SOURCES

EPA Factsheet Number 130, October, 1986.

Exotoxnet, Extension Toxicology Network, Pesticide Information Project, Michigan State University, January, 1989.

EXHIBIT 4 Toxicity and Water Contamination Information for Pesticides Used in Minto-Brown Island Park (Doherty, 1989)

Chemical	Acute				Chronic†					Inadequate environmental data	Water contamination ¶
	Oral LD50‡	Acute Toxicity	Eye/skin effects	Other organ damage	Ter	Mu	Repro	CA	Inadequate chronic data		
Herbicides amitrole	>400	slight	+§	+			+	+	+		
atrazine	1750	slight	+			+	+	+			+
bentazon	2063	slight	+		.		.	.	+	+	+
2,4-D	300-1000	slight to moderate	+	+	+		.	+			+
triclopyr	515-1127	moderate		+	+		+		+		+
dichlobenil	4250	slight							+	+	
dinoseb	20-60	highly		+	+		+	+	+		+
diquat	215-235	moderate	+	+	+	+	+	+			
EPTC	1650	slight	+		.		.		+		+
metolachlor	2780	slight	+	+	.	.		+		+	+
metribuzin	1100-2300	slight	.	+	.		+	+	+	+	+
simazine	>5000	not so toxic	+	+	+	+	+		+		+
trifluralin	>10,000	not so toxic	+	+	+	+	+				+
Insecticides carbaryl	255-800	slight to moderate		+						+	
carbofuran	3.8-35	highly							+		+
disulfoton	2-6.2	highly							+	+	+

	Oral LD$_{50}$	toxicity								
fonfos	3-18.5	highly						+		
permethrin	1500	slight	+	+	·	+	+			
Fungicides benomyl	>10,000	not so toxic	+	+	+	+	+			
copper hydroxide	1000	slight	+						+	
mancozeb	>5000	not so toxic	+	+	+	+	+	+	+	+
metalaxyl	669	slight			·	·	+			+
vinclozolin	>10,000	not so toxic	+		·	+	+		+	
Growth Inhibitor maleic hydrazide	6950	not so toxic		+		·	·	+		+

Note:
† Mu = mutagenic effects; Ter = Teratogenic effects; Reprod = Reproductive effects; CA = Oncogenic/carcinogenic effects.
‡ Oral LD$_{50}$ = oral dose needed to kill 50% of tested animals.
¶ Water contamination = detected in water or expected to be mobile in soil.
§ Yes or positive results in at least one study = +; Negative results in most studies = -.

EXHIBIT 5

EXCERPTS FROM THE EXECUTIVE SUMMARY OF THE PESTICIDE REVIEW FOR

MINTO-BROWN ISLAND PARK (Doherty, 1989)

"Managing pesticide programs had changed dramatically in the 1980's. Neighboring residents and park users may now want to know more about the toxicity and location of pesticides used. In addition, emergency suspensions by EPA and voluntary withdrawals by manufacturers has made many pesticide users wonder how reliable past research data really is. Professional applicators now know that "EPA approved" pesticides can't always be assumed to be "safe" for people or the environment.

". . . . The Audubon Society became increasingly concerned over pesticide use in Minto-Brown Park after EPA focused attention on the toxicity of dinoseb, an herbicide used in the park. In 1986 EPA issued an emergency suspension order stopping use of dinoseb due to its ability to cause male sterility, birth defects and numerous other health and environmental effects.

". . . . Several cities and the state of California have stopped or are proposing bans on the use of pesticides which can cause cancer, birth defects or environmental damage. The city of Salem should take similar action. The Alternatives phase of this project will help to identify effective, less toxic and more environmentally acceptable pest control methods. In light of the data presented in this Pesticide Review and the increasing concern on the part of many park users, the Salem Parks Department would be well advised to adopt a policy of stopping the use of pesticides known to cause long term health or environmental damage."

The Case of the Lecture that Wasn't

By Ronald L. Cherry, Juniata College,
Huntingdon, PA

[Editor's note: This "case" was written two days after the author returned from serving as an invited speaker at a workshop on "Teaching with Decision Cases" held at Michigan State University, August 15-16, 1994. It was prompted by the reaction of workshop participants when Dr. Cherry opted not to lecture about "Teaching Techniques," preferring instead to elicit from participants their own ideas.]

During Monday afternoon of our case workshop, we collectively participated in a live case. Bringing closure to a case, especially a live one, is one of the more difficult choices which a case facilitator has to make. In this particular situation, I have chosen to close the case after the "class" has put the event aside in the rush of activity that goes with the opening of the academic year. In what follows I shall make no effort to bring closure to the process aspects of the case even though they clearly overwhelmed the substantive issues which could have been considered. Rather, I will make an effort to bring closure to the substance of the case, a substance which we were collectively unable to explore in an inductive manner.

The real question of case pedagogy or teaching technique is this: How does the facilitator guide or intervene in the discussion process to create an effective learning environment? The case discussion process presents the facilitator with a series of decision points requiring concrete action. If there is a general model of case pedagogy, then it must help the facilitator navigate these decision points.

In my own experience with the case method I have constructed a set of these decision points or what might be called decision variables confronting the case facilitator. I have grouped them in several categories (see Exhibit 1). If you have had some experience with cases, you have probably accumulated your own list. If you are just beginning to use cases, you will come to develop a list. In either case, your list will vary from mine and you may group

things differently. That is all perfectly legitimate as the specifics of the list constitute only a part of the "model" which I would like to share.

There are differences in these decision variables not only in content, but also in the type or variety of dimensions to them. For example, some may allow a simple yes/no response; such as whether or not to use a mini-lecture (item E.4., Exhibit 1) in the case discussion. Others may present a continuum along which the facilitator may operate, e.g., one can summarize (item G.9., Exhibit 1) a lot or a little, frequently or infrequently. Still others may offer a wide selection of choices which are quite distinct , e.g., the physical arrangements of the setting (item E.8., Exhibit 1).

In any event, the facilitator needs to lay out the range or variety of actions which could be taken with regard to each one of the decision variables. This is an incredibly "long" exercise, and one which each new case experience will tend to modify. An example is found in Exhibit 2 where a range of actions for responding to a dominating participant (item G. 13, Exhibit 1) is listed. The list is by no means exhaustive.

Having constructed a range of interventions which the case facilitator can make, the third step is to assess the likely impact of each of the potential interventions. Here, the process assumes the character of a prediction or hunch, and risks are clearly involved in the quality of the facilitator's predictions. Nonetheless, in Exhibit 3, I have hazarded a guess about some of the outcomes which might occur if I choose to take item 4 from the list in Exhibit 2. These are not far-fetched possibilities; I have had each of them occur. Sometimes they were what I wanted to occur; sometimes they were not. (Yes, it is sometimes useful to have the student try to and be successful in producing a rebellion.) When I got a response which I did not want, it was frequently because I made a poor prediction or lost track of the goals I had in mind.

In order to minimize the number of failures in the intervention process, it is useful to keep a variety of basics in mind. First, try to keep the course objectives and the objectives for each case firmly in place. These objectives point the way to the desirable outcomes that you want created in the situation.

Second, know the territory. This means not only knowing the case thoroughly but also knowing yourself and each student thoroughly. In a required course for all students in our programs at Juniata, we spend a significant amount of time using tests and exercises to help each student build a "personal profile". (This is a term which I much prefer to "personality" as the latter does not as easily allow distinction between being and behavior as does the former.) The Myers-Briggs, TAT, FIRO-B, Parker Team Player Survey and a number of other instruments are used to help the student build his/her profile. Thus, when appropriate, the case facilitator may legitimately ask the student to assess the degree to which his or her position, behavior, etc. reflects his/her profile. Now incidentally, we who teach this course construct a profile for ourselves and make it public for the students. We will frequently point out how our position or behavior is reflective of it and we will also point out when our behavior is not reflective of it and explain why we have consciously chosen to walk an unaccustomed path.

And finally, consider the degree of cohesion within the class, the amount of experience which the class members have had in case studies, the variety of student backgrounds represented in the class and other such givens.

When I said, "It all depends," I was not trying to be flippant.

Look now at the "general model" which I have tried to sketch out:

I look at the variety of choices which I have. For each choice I establish a set of possible interventions. For each intervention I predict the likely outcome, taking into account the givens. I then compare the likely outcomes with those which are desired on the basis of the objectives, and I make a choice.

Can I consciously do all of this each time? More critically, can I consciously do all of this while the case is in progress? Obviously not. But I would suggest that unless I work on this continuously, my facilitator style will cease to be situationally based and will become more of a mechanical drill.

I would like to add before I close that the workshop helped me look in new ways at several issues. First, I became more sensitive to the tug between (1) using the method to "teach" group process skills, and (2) using it to teach decision-making skills for technically sensitive decisions. Second, I am rethinking the selection of cases I make on the basis of the degree to which the case writer chooses to set forth one or more plans of action. And finally, I learned some creative ways to have students hypothesize about data not in the case. A very useful two days for me!

I would like to thank all of you for your work in the sessions. It is clear that with more or less success at this or that point Chris and I were able to "establish the conditions in which learning could take place." That is what we tried to do.

Thanks to you all, and good luck as you set forth on the new or continued use of case studies. I can assure you that it is a process, once launched, which will continually challenge your view of "teaching."

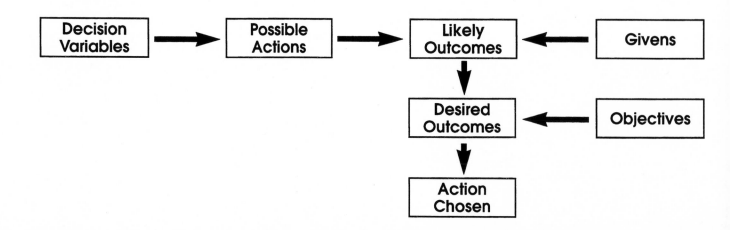

Exhibit 1: Issues/Decision Variables Confronting the Case Facilitator in the Process

A. Roles Sought for the Student

1. Inquirer
2. Hypothesizer
3. Selector
4. Advocate
5. Challenger
6. Compromiser
7. Gatekeeper
8. Listener

B. Student Anxieties Likely to Arise
1. Uncertain about what is expected
2. Desire for single question/single answer
3. Suspicions about cooperative effort
4. Fear of playing an active role
5. Reluctance to change behavior
6. Fear of "learning" as a verb

C. Potential Roles for the Facilitator
1. Challenger
2. Summarizer
3. Facilitator
4. Pacifier
5. Director

D. Faculty Anxieties Likely to Arise
1. Loss of control
2. Worry about process
3. Loss of authority-figure status
4. Worry about degree of preparation
5. Worry about adequacies of skills
6. Worry about being liked
7. Worry about the necessity to "drive to action"
8. Worry about evaluation techniques

E. Mechanics
1. Use of the blackboard
2. Use of role playing
3. Use of small group activity
4. Use of mini-lectures
5. Use of readings/lectures before or after the case
6. Frequency and nature of evaluations
7. Team teaching
8. Physical arrangements
9. Case resource sources

F. Teacher's Expectations for the Student
1. Preparation
2. Use of case evidence
3. Creation of missing data
4. Attendance
5. Attentiveness
7. Creation of a plan of action (POA)
6. Participation

G. "Points of Choice" During the Case
1. Case opening
2. Encourage/limit inquiry at a given level
3. When, how to progress to a new level
4. Encourage/discourage an individual's line of approach
5. Require/finesse case data, especially quantitative data
6. Bring closure to analysis, move to action plans
7. Interrupt with mini-lectures
8. Degree of detail required in a POA
9. Summarize, interpret
10. Challenge, draw out
11. Tidy up/permit confusion
12. Insist on a single POA?
13. Deal with the dominant participators
14. Deal with the non-participators
15. Bring closure to the case
16. Critique the class performance

Exhibit 2: Choices for Dealing with a Dominant Participant (Item G.13., Exhibit 1)

1. Do nothing

2. Choose not to call upon him/her at all

3. Limit the number of opportunities for him/her to enter the discussion

4. Counsel with the individual outside of class

5. Challenge the substance of the student's remarks

6. Instruct the student in class to alter his/her behavior

7. Prohibit the student to speak in class

8. Insert the individual into a role play situation in which the class can legitimately discuss how his/her style impacts the group

9. Counsel with one or more students outside of class in order to set them up as a challenger to the individual leaving them to craft a strategy to "control" the dominator

Exhibit 3: Plausible Outcomes of Strategy No. 4, Exhibit 2

1. Student may understand your point and willingly moderate his/her behavior

2. Student may perceive that he/she is being unfairly singled out

3. Student may be confused about your intervention, especially if participation is a large part of the grade

4. Student may become angry and aggressive, or passively pout

5. Student may set out to lead the class in a rebellion, and so forth. (Each of these reactions then requires a response, etc.)

Thea Grossman

By Sally Lyon, Senior Research Associate,
Center for Case Studies in Education,
Pace University

Thea Grossman was going to try something drastic! For the first five weeks of this Effective Teaching Methods course, Thea had been cajoling, prodding, and occasionally begging her students to participate in class. Convinced by her years of college teaching experience and study of teaching/learning theory and adult development that students would never internalize educational theory if they themselves learned in a passive way, Thea had come to champion participative teaching methods over traditional lecture formats. She felt strongly that her students should be expected to apply theory to actual teaching situations and that class time should be used for analysis and discussion, not lectures. The students could read about the principles of learning in the text; in her classroom, they would learn to use and apply them.

Students at Metropolitan University, however, were generally resistant to her plans; listening to lectures and responding to traditional exams had made them reluctant to think out loud or to expose their opinions for peer review. This class was no exception. The 23 women and six men in the class — all sophomores or juniors with several education courses behind them — were passive in spite of Thea's exhortations to become active learners and her reminders that their participation (or lack of same) would be factored into their grade.

But today would be different. After an exceptionally desultory exchange during Tuesday's class, Thea had made up her mind to "shock" the class into action. And she had the perfect subject to work with. The chapter the students had been assigned for today was on developing communication skills. The textbook compared effective teaching to acting, noting that teachers had to see the classroom as their stage. Other topics included non-verbal communication — using body language, facial expressions and movement, and reading the body language, facial expressions and movement of others, — and emphatic listening.

As Thea entered the classroom on Thursday morning, she began to push all of the furniture against the walls, leaving a large open space in the center of the room.

Thea perched on the edge of a table against the wall and watched the surprised expressions as students entered the room. They reacted with self-conscious giggles and protective poses, clustering as near the outer edges of the room as possible. Thea noticed with some amusement that the students were standing as closely as possible to the position in the room where they usually sat. When the bell rang, Thea began without preamble. "I want each of you to take a name tag and write on it the name of a famous person. Someone you admire." She began handing out blank "Hello, my name is..." tags and magic markers for the students to use. When she noticed that most were not writing and looked confused, she went on, "Stop for a minute and think of someone you might want to be. The only requirement is that the person you choose should be someone everyone else in the class would know."

"Anyone?" giggled Angela.

"Yes, alive or dead, real or fictional. A famous person you admire. It could even be someone of the opposite sex." To make her point, Thea took a name tag, wrote "Jesse Jackson" and pressed it on the lapel of her blazer.

"You mean like, a singer?" Jennifer was clearly confused.

"Sure, if it's a person you admire," replied Thea. "Everyone have a name tag?" OK, now you are to pretend you are that person. Turn to the person next to you and, one at a time, interview each other. Your task is to find out one thing about the person that NO ONE knows. Since you really don't know the person, you'll have to make something up. But you should make up something that could be in keeping with that person's character. Do you understand?"

The students just looked puzzled. Thea smiled and pushed ahead. "OK, for example, someone who interviews me might find out that Jesse Jackson wanted to be a rock star when he was a kid."

"Don't you mean Michael Jackson?" Sonia asked.

Again, Thea smiled. "No, Michael Jackson IS a rock star. Jesse Jackson is a politician."

Sonia just shrugged and a hum of general confusion greeted Thea's explanation. "What do you mean? Like, act?" asked Peter.

"Yes, assume the role of the person you chose and try to talk about that person in character." Thea began to physically indicate partners, as the students remained standing with confused and embarrassed looks on their faces. "Jenny, you talk to Chris...John, you're with Abby...Marie and Beth...Chrissy, you work with Katherine..."

Gradually the students turned to the person next to them, and Sarah heard halting questions of each other: "Uh, what about you do you want me to know?"

"I don't know. I don't know what we're supposed to do."

"Uh, I think you have to tell me something about you that I don't know. Like, make it up."

"Like what?"

"I don't know." Katherine, the student who had begun the exchange, looked at her partner's name tag and said, "Like, make up something about Kathie Lee."

"Like, did you know she's pregnant? Like that?"

"No, not that. She really is pregnant. Everyone knows that."

"Not me," thought Thea. "I don't even now who Kathie Lee is."

Thea moved to another group to observe Sarah and Barbara.

"Do you want to start?" Sarah asked Barbara.

Barbara rolled her eyes and nodded. "You ask me stuff," she said.

"All right, what secrets do you have?"

"Secrets!" cried Barbara, as if the thought of telling secrets about Paula Abdul had never occurred to her.

Sarah went on. "OK, just tell me something about Paula."

Barbara seemed bored. "I think it should be about the music business."

"OK," Sarah said. "Just say something," in a tone that suggested that Barbara's reply would not matter in any case.

Thea continued to circulate and finally interrupted the exercise. "All right – how did you do? Let's go quickly around the room. Each of you should "introduce" your partner by telling the new fact you learned. Katie, why don't you start?"

Katie shrugged and said, "Melissa is Snow White and she has a crush on Sleepy."

Everyone laughed, and Thea began to relax. "OK. Snow White, what did you learn about your partner?"

"Katie is Madonna and she grew up poor in New Jersey."

Several people commented that everyone knew that about Madonna, and Thea moved quickly to the next pair.

"Barbara, who was your partner?"

"Sarah."

Thea tried not to sigh. "Right. But who was Sarah?"

"Oh, she was Christie Brinkley."

"What did you find out about Christine Brinkley?"

Barbara corrected her, "Christie. She wants to have another baby."

"OK. Sarah, who was Barbara, and what did you learn?"

"Paula Abdul and she's going to tour with Michael Jackson."

Thea couldn't think of anything to say about either of the people Barbara and Sarah had chosen, so she turned to the next group. Most of the students had very little to say, and no one had topped Melissa's revelation when Thea came to Dina and Cherie. "Dina, you're next."

Dina looked embarrassed. "Cherie is Vanna White. But she didn't know anything about her to say."

Cherie said, "Yeah, what could I say?"

Thea said, "You needed to make up something, something that might be interesting about Vanna White that no one knows."

Cherie looked puzzled. "Like, how would I know?"

Thea, racking her brain to remember who Vanna White was so she could give an example, responded, "That's the point. You were supposed to create something about your character — her personal life, her background, her likes and dislikes..." Thea interrupted herself. "That's what people have been reporting so far. Don't you understand?"

Cherie shook her head and Thea quickly completed the introductions, anxious to move on before she lost the group altogether. By the time everyone had responded, three students had been unable to think of anything they might know about their choice. Thea wondered if that problem was a reflection of the shallowness and unidimensionality of the people they chose — cartoon characters, TV personalities, models, rock music stars. Not one person had selected someone of substance, someone who would have "secrets" to reveal.

Once everyone had responded, Thea asked if the activity had been fun. A few students responded.

"Sure."

"No."

"That was hard."

Thea acknowledged the general reaction. "So you found it difficult to communicate as someone else. Anyone have any ideas why this wasn't easy to do?"

Again the group ducked and dissembled. Sarah heard murmurs of confusion: "This is stupid."

"I don't get this."

"What is she doing, anyway?"

Thea absorbed the mood of the group and made a mental decision to proceed rather than try to pry further analysis from them.

"All right. Now I want you to find another partner. Someone you don't know too well." Thea herded the students around, trying to steer them toward unlikely partners. "Anita, you work with Chris... Marla and Brian, Beth and Joanie..." When the class was again paired off, she unveiled the next surprise.

"Have any of you done the `mirror' exercise? You and your partner are to copy each other's movements as if you are each other's mirror. Jenny

help me." Thea motioned for a student to leave her partner and come forward. "Move any way you want and watch me copy you." Jenny returned Thea's gaze with little expression and did not move. "Move your arms or something," Thea explained. "Pretend you're looking in a mirror."

Jenny turned her head toward her classmates, clustered around the room and looking at her with expressions of mirth, embarrassment, and glee. She tentatively raised her right hand to her head.

"Right, good!" cried Thea, as she raised her left hand in imitation of the gesture. "See how I am `mirroring' you? Keep moving." Jenny put her right hand down and turned her head to the side. Thea did the same. After a few additional tentative movements, Thea broke it off and turned to the group. "Do you understand? Be each other's mirror. Thanks, Jenny. You can go back to Jennifer. OK, everyone, start." Thea again circulated to watch as students began the exercise, aware of the laughter and hesitant movement around her. She had enjoyed this exercise in theater classes long ago and knew the concentration it took to truly mirror another human being. But reactions she heard around her were flippant and pained. Certainly no one seemed to be concentrating.

After five minutes or so — it seemed like an eternity — Thea interrupted her class' hesitant, embarrassed "dance." "All right, I have a question. Who led and who followed?"

Students looked at Thea dully. "Who led and who was the mirror?" Thea persisted. A few hands were tentatively raised.

"Abby, who led in your group?" asked Thea.

"Uh, I did, I guess. You mean who moved first?"

Thea addressed her reply to the class at large. "Didn't you find that you had to decide who would initiate movement and who would reflect it? How did you decide that?"

"We sort of traded back and forth without talking about it, I guess," volunteered Gail.

"So taking turns was implicit in your interaction with Teri Ann?" Gail and Teri Ann looked at each other and nodded. "How did other pairs decide?" continued Thea.

John raised his hand. "I just started leading and Christine just followed."

"So in another case one team member took control. Did Christine ever take it back?" Christine had half-hidden herself behind Jennifer and was giggling and blushing painfully. John shook his head and responded "No."

"Then this could be about power, couldn't it?" asked Thea, perceiving again that she had better get to the point quickly. "I purposefully did not give you directions ahead of time about leadership or taking turns so that you could see how power relationships and controlling behaviors develop between people almost automatically." A few students' expressions reflected the interest of discovery, but most were busy reshuffling their positions to leave their mirror partners and return to safe territory.

"All right, one more exercise," called Thea over their movement. "Pick a new partner, someone you haven't done something with yet, and I'll explain." Students quickened their pace back to friendly faces and paired off accordingly. Thea assisted the process and instructed, "You and your partner are to interview each other without using words. You are not to talk."

"How can we do an interview without talking?" interjected Jack.

"You'll have to figure it out," Thea smiled. "You are to interview each other about what each thinks of the death penalty: Are you for it or against it? Why? Find that out from your teammate without using words."

"No way!"

"The death penalty — my God!"

"This is totally bizarre." Students snickered and rolled their eyes with incredulity in response to this direction.

"Go ahead – try to communicate without your mouths!" Thea laughed and smiled as she traveled from pair to pair trying to launch her game, but her energy was ebbing; maintaining the momentum with such a recalcitrant group was exhausting. Gradually, the students turned to each other and began, with gestures and facial expressions, to do what they were told.

With fifteen minutes remaining in the period, Thea closed this final exercise. "All right, that's all." She thought that everyone looked relieved. "Get chairs and let's put them in the center of the room in a circle." Students had to pull tables from the

wall to get chairs from behind them, but eventually a circle was formed and the class sat facing each other. Thea felt how different it was to be sitting without a table as protection.

"First, let's poll the group for the results of your interviews. Beth Ann, you start: did you find out Barb's position?"

"Uh, she was for the death penalty, right?" Beth Ann directed her question to Barbara. "Because it says in the Bible an eye for an eye, right?" Barbara nodded verification.

"And what does Beth Ann believe?" Thea asked of Barbara.

"Um...she is for the death penalty, I think, because if you kill someone like you have it coming?" Barbara, too, phrased her reply as a question, seeking Beth's verbal validation of their nonverbal communication.

Beth Ann nodded her agreement. "Yeah — it's not fair to the victims," she elaborated.

"What about you, Kimberly? What were the results of your interview?" Thea proceeded around the circle to query each pair about their non-verbal conversations. Thea was somehow not surprised to hear that 25 of the 26 students present favored the death penalty, mostly for reasons of retribution or based upon Biblical interpretations.

"Was it hard to find this out?" Thea asked.

A choral response greeted her question. "Not as hard as I thought it would be."

"No, not so much."

"It was okay."

"You all seemed to be able to communicate. Did anyone not know how the other person felt?" Thea searched for controversy or debate amidst their collective nods and shrugs and, finding none, decided to switch gears.

"Well, let's debrief the hour as a whole. What do you think was happening in these exercises today?" Thea settled back in her chair, prepared to wait for a reply because of her redirection. But when no one answered after several seconds, Thea felt disappointed. She had expected some reaction to this session, in spite of the class's general reluctance to speak. The activities the students had done had been so unusual she thought they would provoke at least a little discussion.

Thea tried again. "What was happening here, do you think? What were some of your experiences?" Still total silence, punctuated with some uncomfortable shifting in chairs. "How did doing this make your feel?"

Thea waited for several painful seconds. "Do any of you see any relationship between the activities we tried this morning and what it means to communicate as a teacher?" Her gaze moved around the circle; she did not lock eyes with anyone but let her gaze travel slowly enough to allow the students time to reflect on her question. When she had completed the circuit without one comment, she sighed audibly. "OK, I give up. Class dismissed." As the students slowly rose to retrieve their belongings and leave, Thea found herself unable to respond with anything but a curt nod to the students who spoke to her. As the students retreated into the hallway, Thea began moving the tables back into position for the next class. Her depression was palpable.

Part II
New
Decision
Cases

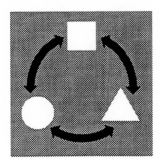

Part II:
New Decision Cases in Agriculture, Food, and Natural Resources

The seven decision cases in Part II were contributed by workshop participants during the three months after the workshop ended. Each case was sent out for peer review to two or three reviewers, most of them conference attendees. All cases required at least minor revisions before being accepted for publication.

In all, eight cases were submitted for inclusion in the proceedings. One was withdrawn, three underwent major revision, and four required only minor revision before acceptance. We thank case authors for their very rapid turnaround time on case revisions and reviews.

We wish to thank Ray William for coordinating review of the "After the Fire" case and the following individuals for contributing their time as case reviewers:

Anita Azarenko, Oregon State University

Josef Broder, University of Georgia

Bernard J. Conlin, University of Minnesota

Colette DePhelps, Washington State University

Steven Ford, Pennsylvania State University

Paul Gessaman, University of Nebraska

William Heald, Pennsylvania State University

Duane Mangold, Iowa State University

Mary Miller, Auburn University

Doris Mold, University of Minnesota

Helene Murray, University of Minnesota

Sherrill Nott, Michigan State University

Claudia Parliament, University of Minnesota

Marla Reicks, University of Minnesota

Andrew Skidmore, Michigan State University

Mary Weidenhoeft, University of Maine

When the Cows Come Home

*By Cheryl J. Wachenheim, Illinois State
University; Scott M. Swinton and J. Roy Black,
Michigan State University*

*"Cattle marketing is an art. It is special not just
because of price variability caused by changes in cattle
supply, beef demand, and factors external to the
industry, but because of the tremendous diversity of the
product; a 1,000 lb. beef animal could be anything from a
heavy steer calf to an overfat heifer going to slaughter.
When marketing is considered as only one component of
operating a feedlot, the manager's task can seem, at
times, overwhelming."*

Craig Schroeder, order buyer

Introduction

Bob Creighton absentmindedly rubbed his chin
as he finished reading a recent research report from
Michigan State University (MSU). What he had just
read, if true, was completely contrary to the
conventional wisdom about the advantages of
keeping a feedlot operating year-round. Bob called
Scott Meyer, the feedlot extension specialist at MSU
and one of the authors of the report he had just finished reading. Scott confirmed the results of the
research; that Michigan farmers feeding one group
of cattle per year could enjoy higher per head
returns than those operating year-round. He
explained that the higher returns were the result of
historically lower feeder cattle prices in the fall and
higher fed-cattle prices in the spring. "Even though
feeding only one group of cattle per year results in
higher per-head production costs because the feedlot is empty for part of the year, the marketing
advantage may be great enough to compensate,"
Scott explained. "However," Scott added, "the
assumptions used may not accurately represent
your situation. You should look at the impact of
seasonal marketing on your own operation. I will
send you some historic price data for Michigan
feeder and slaughter (fed) cattle, but I encourage
you to pull your price data together as well."

As Bob hung up, a multitude of thoughts ran
through his mind. "Even if the numbers say profits
will increase if the shift is made from continuous to
seasonal marketing, what implications does that
have for my decision of whether to sign the
contract with Jennings Beef Packing Company?
And, what would an empty feedlot during part of
the year mean for family members and other
employees working on the farm?"

The Decision Maker

Bob Creighton is a farmer-feeder in southern
Michigan. He has owned and operated a feedlot on
his home farm for 20 years. Neighbors describe Bob
as intelligent, innovative, and friendly. The well-
being of his family and employees are important to
him. His compassion for people, however, does not
diminish his ability to run the farm as a business.
He works hard to instill in his children, Mike and
Lisa, the importance of objective decision making.
He often encourages them to consider how alternative marketing and production strategies will influence the profitability of the business. In fact, Mike
and Lisa each have their own desk in the office and
are frequently included in, or asked for an opinion
on day-to-day and long-term farm decisions.

Bob keeps up-to-date on current technology and
research by reading the popular press and research
reports he regularly receives as a result of his close
relationship with faculty from MSU and his active
participation in the Michigan Cattlemen's
Association. He looks upon the results of the
research report which details advantages to seasonal marketing as evidence of a potential opportunity which should not be ignored. "This is a good
opportunity for Mike and Lisa to learn the
importance of keeping up-to-date on research and
to practice developing a farm plan," Bob decides.
He still regrets not including them two years ago in
his decision to purchase a second farm. He remembers their surprised reaction to his announcement
that he had purchased the farm as a first step in
reaching his long-term goal of increasing the size of
the operation to include them as partners.

Cattle Marketing

Bob Creighton understands the importance of a well planned cattle marketing program. In fact, he considers his feeder cattle buying program as an important factor in the success of his business. Bob buys most of his feeder cattle in the 500 to 650 lb. range from North Carolina and South Carolina where he has an order buyer who consistently sends high quality loads.

Conversely, Bob has never focused on fed cattle marketing because experience has taught him that there is little a producer can do to influence fed price aside from providing high quality cattle. He was, therefore, surprised to learn of the new fed cattle contract offered by Jennings Packing Company in Pennsylvania. Tom Jennings, president of the company, held a producer meeting in Michigan the previous week. He explained how producers could earn a $2 per hundredweight (cwt) premium for cattle slaughtered during the fall months (September, October, and November) by committing, in advance, their fed cattle during these months to Jennings Packing Company. Bob followed up with Tom after the meeting. "As part of the contract, I would like a commitment of not fewer than 500 head of fed cattle during each September, October, and November over the next three years," Tom explained. "In return, you will receive the $2/cwt. premium on all cattle shipped to us during those months. I will send a contract with my buyer in two weeks. It must be signed and returned no later than 14 days from today."

"This opportunity would help guarantee me a break-even price for my cattle," thought Bob. "As such, it would reduce the extra cash flow risk I assumed when I took out the mortgage on the second farm."

Bob has never had a risk management strategy for fed cattle price. Although he recognizes the value of futures, options, and forward pricing contracts for risk reduction, he has never used them to lock-in a price for cattle. He has, instead, counted on his practice of buying and selling cattle year-round to even-out the high and low prices received. He does, however, annually lock-in a price for his grain through his local elevator.

The Farm Facilities

Cattle facilities are on two farms. The home farm has room for approximately 3,200 animals. There are three barns. The "old barn" has both slatted and concrete flooring. Inside the barn, there are two pens with concrete flooring used to start cattle. Each pen can comfortably hold one load of calves (approximately 70 animals). Feeder cattle are also started in the two dirt lots adjacent to this barn. These four starting pens are an important part of the operation. The herdsman likes to keep incoming cattle on solid flooring (concrete or dirt) for a minimum of 10 days. The remaining pen space on the home farm has slatted floors. Therefore, if the starting pens are needed for additional incoming cattle, after 10 days cattle are moved to a pen with slatted flooring but the gate to the alleyway is left open so cattle can move back to a solid floor to rest their legs periodically while they get used to the slats. If cattle do not have 10 days to become acclimated to concrete, they go on feed slower and have more leg problems.

On the second farm, there are two barns that, together, hold approximately 2,000 animals. Bob purchased this farm two years ago. It is seven miles from the home farm. All the flooring in both barns is concrete. Although the barns were in general disrepair when purchased, they have been repaired and well maintained since Bob has owned them. One of the barns has a good machine shop and an office.

Farm Employees and Their Duties

In addition to Bob and his children, there are five full-time employees on the farm. Salaries range from $7 to $12 per hour, high enough to attract and keep good employees. In addition, each employee receives $40 per month towards health and dental expenses.

The herdsman for the home farm, Fred, has worked for Bob for 10 years. In addition to managing the day-to-day activities, Fred frequently participates in farm decision making. Bob recognizes the value of his input. For example, several years ago, Fred was an active participant in designing a new barn on the home farm. Several of his ideas about facility layout have been tremendous time savers. Like Bob, Fred frequently pushes a pencil when a decision has to be made. In doing so, he is able to support his suggestions by showing how they will impact the profitability of the business. A second full-time employee works with Fred on the home farm. Together, it is their responsibility to take care of the cattle and facilities. They load and unload feeder and fed cattle, sort fed cattle, walk pens daily, precondition incoming cattle, treat sick cattle, and repair and maintain facilities.

A third full-time employee divides his time between feeding cattle on the home farm and helping to scrape and haul manure on the second farm. Two other employees work on the second farm, including a herdsman. The herdsman walks pens daily, checking and treating for illness or injury. He is assisted when loading and unloading cattle, occasionally when treating sick animals, and when sorting cattle. He also keeps the herd records and is in charge of ordering supplies and feed. The duties of the third man working on this farm include feeding, which is done twice a day, and helping scrape and haul manure, done weekly in each concrete floored pen.

The remaining employees include Bob's son and daughter and Earl, a long time employee. Mike and Lisa Creighton returned to work full time on the home farm nearly a year ago. Mike recently graduated from Central Michigan University with a bachelors degree in business. He will be getting married next October and plans to continue working on the farm. His interests lie in farm machinery. In fact, he spends so much time in the machine shop, Bob provided him with a second desk there. "The machine shop is a key factor in the success of our business," Bob told Mike as he showed him his new desk. "Equipment maintenance and our long term practice of purchasing and retrofitting used equipment to fit our needs contribute to the profitability of the operation." Mike shares the machine shop with Earl, who has worked nearly full time on the Creighton farm since 1980. Over the last year, Earl, who is nearing retirement age, has cut down to about 20 hours per week.

Effects of Seasonal Marketing on Labor

Bob has always been willing to pay his employees a fair wage because he believes good employees are one of the keys to success in the cattle business. While he contemplates a change to seasonal marketing, Bob finds himself continually returning to how it would change his labor needs and cost. "I ought to be able to custom-hire some of the chores during the busy harvest season if seasonal marketing requires that I get by with fewer full-time employees and I cannot find good seasonal help," thought Bob. "Planting corn is already custom-hired. I could custom-hire chopping corn for silage and spreading manure. However, because we are set up to purchase and retrofit used machinery and equipment, it is important to keep the machine shop fully employed. In addition, custom-hiring a job will cost more and seasonal hiring will result in longer hours to do the same work as workers become (re)acclimated to the farm routine." For a fleeting instant, the idea of opening the shop up to work on other farmer's equipment entered Bob's mind. He quickly discounted that idea because other farmers will need their machinery worked on at the same time that he will need the extra labor in the field.

Bob also considered the effect of a seasonal marketing plan on his crop enterprise. "Because my brother does the planting, feeding more cattle into the spring would not create as much of a strain on existing labor as it would for other feedlots in the area. However, corn silage harvest in the fall is the busiest time of the year for us. Bringing several thousand head of cattle into the feedlot in the fall under a seasonal marketing plan would force me to layoff some of my current employees, or move them to part time and hire more part-time help for the feedlot during this period. Unfortunately, this period coincides with harvest season when these part-time employees could work for other farmers in the area to supplement their income."

The Crop Enterprise and Feeding Program

The Creightons plant about 700 acres of corn for silage, all of which is fed. Bob's brother has traditionally done all the planting on a custom-hire basis. Over the past year, Mike has expressed an interest in taking over this responsibility. Although corn silage provides most of the roughage used in his operation, Bob buys the hay used in his starter diet, all his grain, and byproduct feeds.

Bob's feeding program is similar to that of other producers in the area, although he feeds more roughage than feedlots in the Central and Southern Plains states. As a result, his average daily gain of 2.8 lb. is comparable to other operators in Michigan but lower than the plains states' average. Recently, Bob explained the importance of the use of byproduct feeds in the operation to Lisa. "Because our rations are flexible enough to incorporate byproduct feeds when available, and in the form available, our cost of gain is lower than it otherwise might be. This feed cost advantage is not just luck. We work with byproduct suppliers to make the farm an attractive customer. Good examples are our practices of taking brewers' grains during the three-month summer brewing season, and sugar (marshmallows and caramel) whenever it is delivered. Recall that we invested in the new Bobcat to move the large containers the sugar is delivered in. We also buy all of our corn as corn screenings from the local elevator. The elevator works with us because we have bought their screenings for several years and our feed storage facilities allow us to take the bulk of it when it is convenient for the elevator to ship it. Seasonal marketing of cattle will reduce our flexibility to purchase byproduct feeds and corn screenings because we cannot store everything delivered if the feedlot is empty for several months of the year."

Seasonality of Cattle Prices

Researchers who developed the report Bob just read used two price series to estimate gross margins on cattle. A weekly price series from the Lexington, Kentucky feeder cattle auction market was used to represent feeder cattle prices for Michigan producers. The use of this series was justified by researchers because most Michigan cattle feeders buy cattle from the southeastern United States. From analysis of this series, researchers found that feeder cattle prices have pronounced seasonal patterns. Upon reading this, Bob consulted the cattle production and marketing handbook provided by MSU Extension. It indicated that seasonal patterns in feeder cattle prices exist, but are "heavily influenced by shifts in the profitability of feeding cattle and by seasonal shortages." This variability in seasonal price trends was not mentioned in the research report. Bob wondered why MSU researchers ignored this important consideration.

To represent fed price, the researchers used an average monthly fed price for beef cattle sold through Michigan Livestock Exchange (MLSE), a prominent livestock marketing cooperative. As with feeder cattle price, Bob's handbook agrees with the research report that there is also a pronounced seasonality in fed cattle prices, but again emphasizes that seasonality in any given year can vary greatly from this long run seasonal pattern.[1]

Bob figured he receives a higher average price for his cattle than producers marketing through MLSE because he does not pay a commission to the buyer, hires a local trucker, and uses his own livestock trailer. He did not, however, think that the difference between his fed price and that offered by MLSE differed by season so he didn't see why he should go back through all his old records and calculate his own average fed price. He knows from talking with packers and MLSE order buyers in the area that his cattle grade and yield are similar to those of producers who regularly sell cattle through MLSE. At least they are for those whose fed prices make up the 1984 to 1992 price series used by researchers to represent Michigan fed price in the report. Besides, he was able to get the average Michigan fed price by month for 1981 to 1992 and for 1984 to 1992 from Bobby Jessee at MLSE (Exhibit 2). In his note accompanying the data, Jessee mentioned that seasonality of fed cattle prices in Michigan and nearby states differs from that found elsewhere in the country because there are more farmer feeders in the region. "Farmer feeders," Jessee explained, "buy and sell fewer cattle during the spring and fall months when they are in the field. As a result, regional packers need to offer higher prices to obtain cattle during these periods." "This is less true," Jessee explained, "in major cattle feeding states in the Central and Southern Plains where most feed is purchased off the farm."

Bob is comfortable that he has all the important numbers to calculate the effect of seasonal marketing on cattle prices, but is not sure which price series to use. Bob recalled a discussion with Herbert Janson, MSU livestock marketing

---

[1] Figures in the research report depicting price seasonality for light and heavy feeder cattle and fed cattle, and figures from Bob's handbook depicting variability in seasonal price trends between years are found in Exhibit 1.

Extension economist. Dr. Janson said that average price should be calculated over the last cattle cycle to reflect highs and lows associated with periods of herd expansion and liquidation. "Cattle cycles," Dr. Janson said, "are about 9 to 12 years in length." However, Bob's handbook says that, like seasonal price differences, length of cattle cycle is quite variable because it depends on conditions outside the cattle industry like the supply of other meats. It seems, from the graph shown in the handbook that 12 years is about right for the last cattle cycle (Exhibit 3). Bob wondered why the researchers only used a nine-year price series when Jessee could have provided them with average fed prices for the last 12 years?

Production Cost

Bob figured that a change to seasonal marketing would also affect production cost because his feedlot would be empty part of the year. Bob's production cost is about $46 and $48 per cwt. gain when the feedlot is 80 percent full year-round and cattle are put on feed at 500 and 650 lb., respectively. From some initial calculations, Bob estimated that the cost per cwt. gain would rise to $48 and $52 per cwt. gain if he only fed one group of cattle on a seasonal basis started at 500 and 650 lb., respectively.

A Time for Action

Bob asked Mike, Lisa and Fred to come over at 3:00 p.m. the next Friday. When everyone was seated around the dining table, Bob said, "I've been thinking about how we raise and market our cattle. Jennings Pack is offering a new contract for cattle marketed in the fall. They'll give a $2 per cwt. price premium. But if we go with them, we'll have to be able to market 500 animals per month in the fall. That means we can't switch to the kind of seasonal marketing suggested in that MSU report. You are all an important part of the business and its future, so I want to talk this through with you and see what you think. Jennings wants a three-year commitment, and we would have to sign-on next week."

EXHIBIT 1

Seasonality and Variability in Feeder and Fed Cattle Prices.

Seasonality of feeder cattle prices

500 to 600 lb. and 700 to 800 lb. steer calves; Lexington, KY; 1984 to 1992.

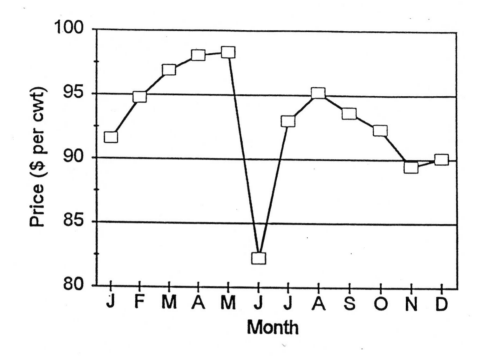

Seasonality of fed cattle prices; Michigan; 1984 to 1992.

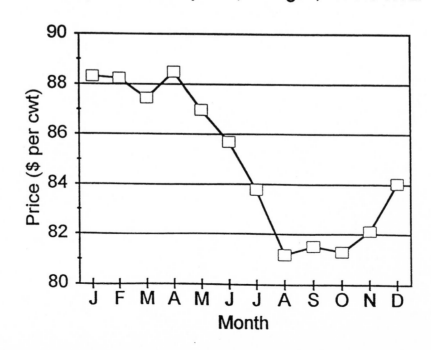

Index of monthly variations in choice steer prices; Omaha; 1980, 1985, and 1987.

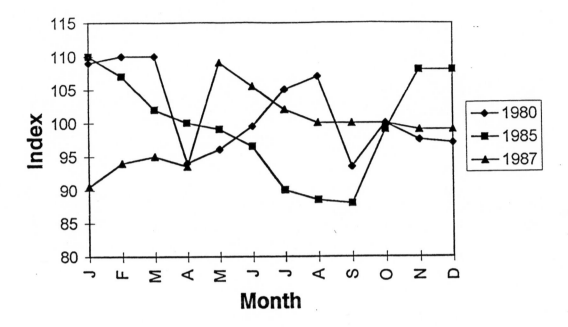

Index of monthly changes in feeder calf prices; Kansas City; 1980, 1985, and 1987.

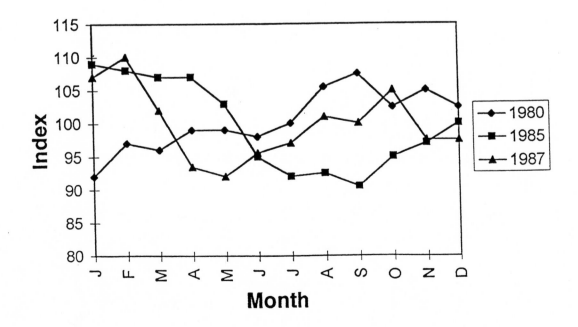

EXHIBIT 2
Average Real Feeder and Fed Cattle Prices[2],[3]

Month	1981 to 1992				1984 to 1992			
Weight	5-600 lb	6-700 lb	7-800 lb	Fed weight	5-600 lb	6-700 lb	7-800 lb	Fed weight
January	93.22	89.75	87.22	91.63	91.58	87.76	85.27	88.32
February	95.92	90.91	88.41	91.58	94.74	88.88	86.16	88.25
March	98.55	91.47	88.04	91.08	96.90	89.18	85.40	87.45
April	99.79	91.86	87.25	91.79	98.05	90.02	84.62	88.50
May	99.18	91.37	85.90	89.96	98.31	90.04	83.64	86.97
June	86.86	81.25	77.57	88.35	82.28	76.27	72.48	85.70
July	93.27	86.86	84.71	86.64	92.96	85.01	83.05	83.77
August	94.24	89.12	85.51	84.05	95.19	88.65	84.99	81.18
September	92.59	87.61	84.80	84.11	93.61	87.69	84.39	81.52
October	90.96	85.82	83.17	83.22	92.31	86.28	83.54	81.29
November	88.98	85.09	93.12	84.04	89.42	85.38	83.64	82.12
December	89.30	85.75	93.88	85.47	90.07	85.82	84.13	84.03
Average	95.71	88.07	84.97	87.66	92.95	86.75	83.44	84.92

2 Prices are inflation-adjusted to a 1992 price level using the Consumer Price Index.

3 Fed cattle prices are FOB farm. That is, the value represents the actual price received by the farmer "at the farm gate."

EXHIBIT 3

Cattle Inventory Cycles

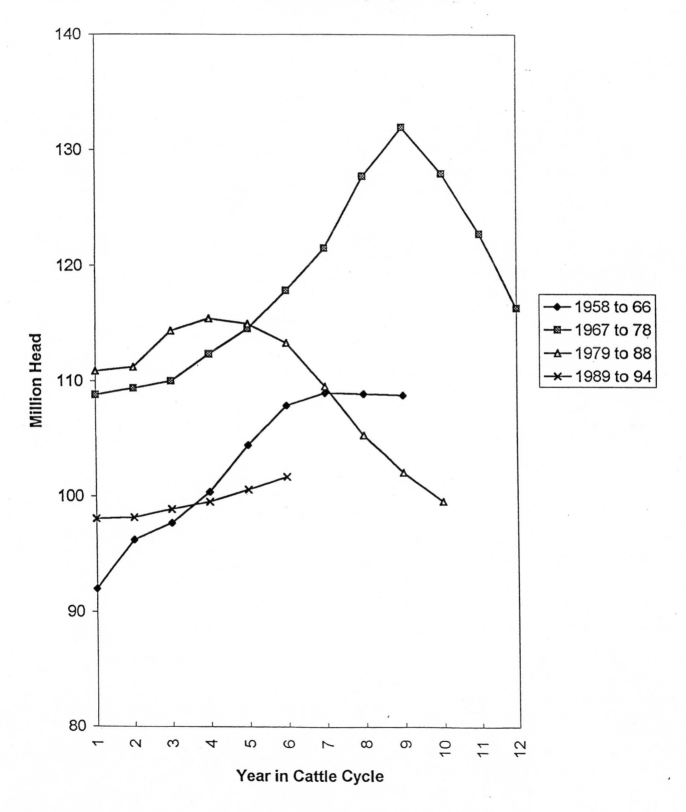

Year in Cattle Cycle

Quality and Coordination in the Food System: The Case of Riceland Foods

By Bert Greenwalt, Arkansas State University; David B. Schweikhardt, Michigan State University; Mary Helen Fairley, USDA Cotton Division, Memphis, Tennessee

Riceland Foods is a farmer-owned marketing and processing cooperative. Founded in 1921 to market rice, it currently markets rice, soybeans, and other grains produced by more than 10,000 farmer members in Arkansas, Mississippi, and Missouri. Riceland is the largest rice miller in the United States, one of the nation's 10 largest grain storage firms, and a "Fortune 500" company. Each year, its 1,850 employees store, transport, process and market more than 100 million bu. of grain. Six mills and two parboiling plants process 60 to 70 million bu. of rice each year, about 20 percent of the U.S. crop. Riceland's business lines include the consumer package market, food service industries, food ingredients, and exports of value-added and raw commodities. In 1993, Riceland marketed its grain and processed products across the U.S. and in 88 foreign countries.

The milled rice produced by Riceland is marketed directly to consumers and food service institutions under the Riceland brand and other private labels. Rice bran and rice flour are sold to food manufacturers as ingredients for use in a wide variety of snack foods, bakery products, cereals, baby foods, and pet foods.

Riceland also processes soybeans into oil, meal, and lecithin (a dietary supplement and ingredient for cosmetics) and merchandises corn, wheat, milo and oats. Soybean oil is marketed directly to consumers and food service institutions under Riceland's Chef-Way brand and several private labels. Soybean meal is an important feed ingredient used by the region's poultry industry.

As a cooperative, Riceland is owned by the farmers whose grain it markets. Each farmer's share of the firm's permanent capital is determined by the volume of grain the member delivers to Riceland. The majority of rice received by Riceland is sold using a marketing pool system. In the marketing pool, each member's price is determined by the average price of all grain processed and marketed by Riceland. Riceland is managed by a 37-member board of directors elected from among its farmer members. This board is structured to assure regional representation among Riceland's members and hires the firm's president and chief executive officer.

Riceland's Market Position

Riceland has long occupied a market position of supplying high-quality and value-added products. While larger grain companies supply products in large volume at low prices and with low profit margins, Riceland has focused on being a quality supplier of niche markets.

Riceland's customers demand rice with the specific characteristics needed for processing rice into consumer products. A cereal manufacturer, one of Riceland's largest customers, has expressed concern that some rice is not meeting the specifications required to produce rice cereals of consistent quality. Because the cereal manufacturer is committed to marketing a "global image" cereal product with consistent quality in all U.S. and foreign markets, consistency of product quality is viewed as a high priority.

The cereal manufacturer maintains that its philosophy is simple: the end user of its products, food consumers, are the ultimate customers. Everyone in the supply chain, including firms such as Riceland, must be responsive to the demands of the final consumer. Its executives cite examples of successful partnerships between itself and its suppliers. A cereal box manufacturer, for example, maintains offices in the cereal manufacturer's plants and provides personnel to work with plant employees in solving problems that arise in the use of its boxes. A California rice supplier was noted for having sent employees to deal with problems that arose in the processing of rice from California. Similarly, a corn supplier has begun to duplicate the quality tests used by the cereal manufacturer on cereal products produced in plants that use the supplier's corn. The cereal manufacturer views itself as pragmatic — it does not require that suppliers meet specific quality standards (e.g., ISO 9000) but instead prefers to work with suppliers to solve problems as they arise. In the cereal manufacturer's view, the common element in these relationships is that the supplier is attempting to be proactive in addressing problems that arise in the cereal manufacturing process.

The cereal manufacturer believes strongly, however, that suppliers of farm products should provide a service rather than merely attempt to sell raw commodities at the lowest price. Consequently, the manufacturer believes that the best relationships with suppliers are those in which the supplier provides a high quality product and effective service at a competitive price. In return, the cereal manufacturer provides a consistent source of business for the supplier. In the cereal manufacturer's view, there is no reason to pay a higher price for improvements in quality — such working relationships are a natural evolution in the supplier-manufacturer relationship in a modern food system. Moreover, improvements in quality management can actually benefit the supplier if such efforts lead to cost avoidance (e.g., fewer rejections of products arriving at the manufacturer's plants). As part of this process, the cereal manufacturer has gone from a philosophy five years earlier of "never telling" its suppliers of its needs and problems to its current philosophy of "telling all" to suppliers in an attempt to find solutions to quality problems.

Events Triggering Problems with Rice Quality

Two events have lead to problems with the quality of the rice shipped to the cereal manufacturer by Riceland. First, the cereal manufacturer installed new automated equipment in its cereal plants. This equipment has eliminated a "touchy/feely" production approach that required operators to adjust the cooking and toasting of each batch of rice. The automated equipment now controls the cooking and toasting of the cereal. Such equipment

allows the manufacturer to produce its global-image cereal products with greater consistency but requires that rice be of consistent quality.

The second event was the introduction of a new variety of rice, Orion, in the southern rice-producing states. Most medium-grain produced by farmers who market rice through Riceland is Orion or Mars. Members of Riceland have begun planting more Orion and reducing their plantings of Mars because of Orion's superior agronomic characteristics. Orion's straw strength is greater, making it easier to harvest. Orion also provides somewhat higher yields than Mars, providing an opportunity for greater profits for farmers. At the present time, Riceland pays the same price to farmers who deliver Mars and Orion rice to its elevators.

After receiving shipments of Orion, however, the cereal manufacturer discovered that the variation in the kernel size of the Orion variety far exceeds the limits that must be met for consistent production of cereal. Consistency in the size of rice grains is critical in guaranteeing consistency in the production of cereal. If the size of the rice kernels varies widely, smaller grains will be overcooked and overtoasted in the manufacturing process, while larger grains will be undercooked and undertoasted. This problem is accentuated by the fact that the new automation equipment does not allow each batch of rice to be adjusted as it is being processed. Consequently, the cereal manufacturer has indicated that Orion is unacceptable. The Mars variety produces kernels that are more uniform than Orion. Mars is acceptable in the cereal production process.

In addition to its preference for Mars, the cereal manufacturer has expressed concern about the random mixing of varieties that can occur as the rice is harvested and transported from the farm to the elevator, and at the elevator. For example, if a load of Mars rice is hauled in a truck after a load of Orion, some Orion may get mixed into the Mars if the bed of the truck is not cleaned between loads. Similarly, the two varieties of rice could be mixed at the elevator if the grain handling equipment is not completely cleaned. Any mixing of the two varieties introduces variation into the rice delivered to the cereal manufacturer.

California is the other major rice growing region in the U.S. The CalRose variety of rice grown in California produces kernels of consistent size that meet the cereal manufacturer's needs. Although the cereal manufacturer had not used CalRose in the past, it discovered that CalRose produces a consistent quality of cereal and now uses only CalRose in several of it plants. The CalRose variety was developed under a breeding program established by the California Rice Industry Council, a trade organization representing all parts of the California rice industry. CalRose is not suitable for planting in Riceland's production regions.

Riceland has provided samples of "mechanically sized" rice (rice that has been sorted using mechanical sorting equipment) to the cereal manufacturer. The manufacturer has indicated that mechanically sorted rice does meet its size specifications and produces a superior product. However, mechanical sorting of rice creates two problems for Riceland. First, the cereal manufacturer has not yet expressed a willingness to pay the additional cost of sorting rice. Second, if Riceland sorts the rice for the cereal manufacturer, it has few outlets to sell the large- and small-size kernels that are sorted out of the rice.

Other Outlets for the Rice

A major brewer has also expressed a preference for specific rice varieties that yield superior tasting beers. Though the brewery has been willing to purchase the Mars variety, it is not the brewer's first choice of varieties. The brewery has expressed a preference for other varieties.

A major baby food processor also buys rice flour from Riceland for use in baby food products. This company has indicated that it is adopting a "zero-defects" approach in purchasing its ingredients for baby food, and that Riceland's rice flour must meet this zero-defects standard. In response, Riceland invested in metal detectors and magnets to prevent any metal from entering its rice flour and has undertaken an aggressive education program using videotapes and other materials to educate its members and employees on the importance of meeting this zero-defects standard. Riceland's 1993 annual report reminded its members of the importance of achieving its quality objectives in its rice flour line:

> "One of the principal uses of rice flour is in the manufacturing of baby foods. There is no room for mistakes in such a sensitive business."

One major advantage that Riceland has in serving the baby food market is that it can assure its customers of the origin of its products. Unlike larger grain companies that purchase and mix grain from many sources, Riceland has control of all of its products from the time they are delivered at its elevators until the grain or grain products are delivered at the customer's location. Much of Riceland's reputation in the market is based on its ability to guarantee the origin and quality of its products to customers.

Meeting the Quality Demands of Its Customers

Riceland has begun to consider alternatives for meeting the quality demands expressed by its customers. First, Riceland recognizes that new varieties of rice are needed to meet the needs of its customers. Rice research is conducted at agricultural colleges using a farmer "check-off" program that collects research funds from each farmer when rice is delivered to the elevator. The use of these funds is directed by a committee of rice farmers. In the past, most of this research has been directed at increasing rice yields or reducing diseases or pests that damage rice during the growing season. Riceland is now attempting to influence the research process by educating plant breeders about the specific needs of food processors (e.g., Riceland, the plant breeders and the cereal manufacturer met to discuss the problems of kernel size). It has formed a southern rice industry advisory committee consisting of Riceland executives and farmers.

Riceland is also considering the establishment of a mandatory varieties list that would specify the varieties of rice its members could plant. Because it is a cooperative business, Riceland has always sought to maintain a "farmer friendly" approach and has accepted all varieties of rice grown by its members. In 1993, over 20 varieties of rice were delivered by Riceland's members. Allowing such a large number of varieties increases the cost of handling rice at Riceland's elevators. Some varieties may also have limited uses if they do not have the product characteristics demanded by buyers. In response to this problem, Riceland issued a "recommended" varieties list in 1993. Farmers were urged, but not required to plant the varieties on the list. They did not receive a price premium for doing so.

Riceland's board of directors is considering issuing a mandatory varieties list, but is concerned that members may object to such a list.

Riceland could also create a price differential by paying a higher price for Mars than Orion. Elevator managers at Riceland have expressed concern that such a system could actually worsen the mixing problem. Since there is no means of determining whether the rice delivered by farmers is Mars or Orion, the introduction of price differentials could create an incentive for farmers to deliver Orion rice under the Mars name and receive the higher Mars price.

Riceland's president has led an effort to communicate the quality demands of customers to his board of directors and members, but past experiences have left doubt about some customers' commitment to the partnership — needed to justify quality improvement programs. A major soup manufacturer, for example, requested a specific type of rice that would maintain its form and texture when used in the production of soups. Riceland made a commitment to become a certified supplier of such rice, only to learn that the soup manufacturer had turned to a lower-priced supplier that, in Riceland's view, had failed to make the commitments required of Riceland. Given this experience, Riceland's president believes that executives both of Riceland and food manufacturing firms must meet to establish a commitment to a long-term relationship that will justify the necessary quality improvements. The philosophy of this relationship must then be communicated to employees at both Riceland and the manufacturing firm if quality problems are to be resolved and commitments to improved quality are to be justified.

Riceland's president also knows that some of his farmer members may resist quality management programs that are not rewarded with higher prices. While some members believe that Riceland's customers are demanding higher quality without expressing a willingness to pay a higher price, the president and board of directors believe the firm must respond to its customers' demand for quality, regardless of whether a higher price is received.

Key Questions for Riceland

Several questions must be considered as Riceland considers its options for dealing with its customers' demands for specific quality characteristics:

1. What obstacles does Riceland face in meeting the demands of customers?

2. What are the consequences of failing to address the customers' requests?

3. What alternatives does Riceland have to meet the cereal manufacturers' demand for rice with a particular size consistency? What are the advantages and disadvantages of each alternative from Riceland's perspective?

4. What are the strengths and weaknesses of Riceland's cooperative organization structure in meeting the needs of its customers?

5. Given Riceland's past experience of attempting to meet the demands of customers, only to lose those customers to other lower-cost suppliers who did not make the investments, what could Riceland do to insure long-term relationships with its customers?

Study Questions

1. What forces in the market can explain why food manufacturers are requesting rice and rice products with specific characteristics from Riceland? (Consider Barkema, Drabenstott and Welch's forces that are causing the "quiet revolution" in the food system.)

2. How does this case demonstrate the increasing need for vertical coordination in the food system? What alternatives are there in this case for achieving such coordination? What are the consequences of these options for Riceland and Riceland's farmer members?

3. The effort to improve quality and serve consumers with high quality food products began at the manufacturer's level and is now trend likely to continue toward the farm level? Why or why not? What forces might cause this trend to continue even further to the farm input level?

4. How might a "service" strategy of selling food ingredients differ from a "commodity selling" strategy? Which is more likely to prevail given future trends in food markets?

Reference
Barkema, A., M. Drabenstott, and K. Welch. "The Quiet Revolution in the U.S. Food Market." Federal Reserve Bank of Kansas City *Economic Review* (May/June 1991), pp. 25-39.

After the Fire

*By Scott M. Swinton and Sherrill B. Nott,
Michigan State University*

The half-seared maple tree reaching over the charred dairy barn summed up the state of the Phoenix Farm. The crops looked green and vigorous, but since the midday fire on June 7, 1994, the dairy half of the business was very much at risk. Things could have been worse. Bill's father noticed smoke from the dairy parlor soon after the fire started. The family saved all but one of their 130 milking cows. The dry cows and replacements were safe on the other side of the farmstead.

Within hours, the Larsons had agreed to make their dry cow barn and milking parlor available for the Phoenix herd. Neighbors helped move the cows to the Larsons in time for the evening milking.

An agreement was made whereby the cows would leave the Larsons' farm by November 1. In July, Bill asked a consultant to review the situation and recommend what type and size of a system they should rebuild. The deadline seemed to be approaching quickly.

The People and the Farm

Grandfather Luke started the original farm. At age 82, he owned a major share of the land, kept the checkbook, and prepared everyone's tax returns. Vern Sr., the father, owned most of the rest of the real estate and ran the crop operation. Son, Emory, was 30; he had a two-year degree in livestock management. Son, Bill, was 28; he had a two-year degree in mechanics and diesel repair. Vern Jr. was 24; he had a B.S. with joint majors in electrical engineering and computer science. All had spouses who made varying inputs into the farm. Luke, Vern Sr., Emory and Bill farmed in partnership, but without any written partnership agreement.

Luke had started with next to nothing. By 1982, when Emory and Bill were invited into the partnership, Luke and Vern Sr. owned 2,200 acres in

central Michigan, 1,800 tillable. Acreage was used for corn, wheat, beans and hay. There was a full line of field equipment, adequate livestock facilities and no debt.

Faced with raising a family of six, Vern Sr. had built up a diversified livestock operation, including 120 dairy cows, during the 1960's and 1970's. But by the early 1970's, he was "all milked out." He began switching over to beef and sheep. By 1982, the only livestock were 40 cows, 25 beef cattle (Holstein, Hereford and Angus), a few hogs and 200 sheep.

The entire livestock side of the business was left to Emory and Bill after they finished up school and joined the partnership. They used their knowledge to rebuild the dairy herd quickly and phase out all the other livestock. They bought registered Holsteins and used artificial insemination as they got the herd up to 130 milk cows, 35 dry cows, replacements, plus 50 dairy steers just before the fire. Their results are summarized in the DHIA report, Exhibit 1. Emory kept a close eye on the cows. Bill installed a total, mixed-ration feeding system that had helped build milk production from 40 lb. to 57 lb. per cow per day at the time of the fire.

The fire ruined the milking parlor, but not all of the equipment room. The bulk tank was salvaged. Aside from the holding area, the barns escaped major damage. The feed was mixed at home and trucked over to the Larson farm daily for the milking herd.

Interviews with the Partners

The consultant came by the last week in July to make arrangements to meet with the partners. Luke declined to meet with him, saying the decision was up to his two grandsons. Emory and Bill preferred to meet separately with the consultant over the following two weeks.

Before meeting with Emory, the consultant talked briefly with Vern, Sr. He was busy delivering last year's corn to fill a marketing contract, so the conversation was frequently interrupted as he moved between the trailer truck and keeping the auger clear. Vern, Sr. was proud of what he and Luke had accomplished, but he no longer wanted any part of the livestock. "I'd rather forget about the financial disaster the sheep caused," he said. He thought the boys had done well with the milk cows in recent years. He was leaving it up to them to decide what to do in response to the fire.

Emory: It's a Job Like Any Other

The consultant talked with Emory while he and the part-time hired help were milking. Emory favored rebuilding a facility the same size as the old barn. For him, the issues were whether to sell the milking cows and heifers so as to rebuild the old barn, or to use that facility as a temporary holding area while building a new barn and refurbishing the old double-six milking parlor. Either way, he preferred to contract out the construction work. His share of the milking and caring for cows was more than enough work for the husband of a professional nurse and father of a three-year-old.

Emory felt farming ought to be a job like any other. A farmer ought to be able to limit the hours worked and take some vacation. Expanding the operation to 300 cows, as Bill wanted, would require two to three employees and closer management communication among the partners. It seemed impossible to find capable, dedicated workers for the dairy operation; they either didn't know how to handle cows or they didn't stick around. The hired man who shared morning milking with Emory at the Larson place was a lucky find, Emory felt, but who knew how long he would stay?

Besides, Emory felt it was hard enough for partners to communicate in a dairy operation their current size. "I've noticed that the most successful farmers have meetings," Emory observed, "they run it like a business." Emory couldn't remember the last time the four partners sat down at a table and planned the business together. But he did remember coming to the machinery shed one morning and finding that big field cultivator Bill bought on the spur of the moment at an auction. And he remembered the time last fall when Bill took a three-day vacation without letting him know ahead of time.

Aside from the problems of managing a bigger dairy herd, Emory felt the family didn't need the money anyway. The 165-cow dairy operation was already big enough to support the families of the two brothers—in fact, it supported Vern, Jr. and his wife as well. And new investments were always risky. A much safer bet was a salary like the one Emory's wife brought home as a registered nurse.

Although she came from a family farm herself, Emory's wife questioned whether the effort he put into the farm was worthwhile. The stresses of getting along with everyone—Bill and his wife in particular—took a lot of the fun out of farming. It just didn't seem fair that Bill could take partnership money and invest it in more farm equipment when Emory couldn't take money to do the things he wanted, like renovate his home or take a vacation. For that reason, Emory had proposed a fairer way to share the partnership's profits: at year-end, profits would be split equally among the partners, and each could do as he pleased with the money. Those who wanted to spend it on the business could do so, those who didn't wouldn't be obliged to. That would surely be fairer than reinvesting every dollar in the business after partner salaries and other expenses were paid.

Bill: An Opportunity to Expand

On another visit, the consultant talked with Bill and Vern Jr. while they took their turn running the parlor. Bill saw the fire's destruction and the insurance payout as an opportunity to expand by building a state-of-the-art 300-cow barn and double-10 dairy parlor — perhaps even a 40-cow Rotoflo parlor. Rather than contract out the whole rebuilding job, Bill thought the family could save money (and build a bigger facility) by being their own general contractor, hiring Amish laborers and working alongside them.

Bill's temperament was the opposite of Emory's in many ways. Bill loved farming and dreamed of building the Phoenix Farm into the biggest dairy in the state. His wife shared his enthusiasm and, despite child care and her job as a licensed practical nurse, she still found time to help out when needed with payroll, animal records, dehorning and castrating tasks. Bill's enthusiasm and fondness for big-picture thinking had made his ideas the center of the family debate over what to do with the dairy. He had seen a 40-cow parlor that would minimize the need for labor in a 300-cow dairy operation. And it certainly looked as though that farming adage was true: "Get big or get out." Bill knew which he wanted. But he also knew the partnership couldn't manage an operation like that in its current state. He was still puzzled why Emory had refused to milk one morning. And he wondered why no one told him the hay was running out before the day it was all gone!

To bring the farm into the 21st century, Bill wanted someone who could understand new technologies, do electrical repairs, manage the books when Luke chose to pass them on and, above all, someone who could communicate with everyone else. Vern, Jr. was the only solution. As they milked together evenings, Bill would try to persuade Vern Jr. to join the business as a partner. "If Vern Jr. goes, I go," he said.

Vern Jr. did not take a position on the dairy barn reconstruction. He had done systems analysis work for a firm in the county upon graduation from college. He quit there and went to work for the farm full time the previous February. Now, two years after graduation, he felt maybe the time had come to look for a professional job that used his computer training. Although part of him really enjoyed farming, it wasn't easy in this family business. "You know the story of the Tower of Babel?" asked Vern Jr. "That's our family."

Preparing for the Next Step

In thinking about whether and how to rebuild the dairy facility, the one thing both Emory and Bill could agree upon was that the new setup had to provide the cash flow and net returns to support four or five families. Since the consultant lacked more detailed farm records, the best planning tool seemed to be average performances of other farms (Exhibit 2). The Phoenix dairy enterprise was a complement to the crop enterprise, adding considerable value to the feed crops by "running them through the cows." While dairy farming had its ups and downs, milk prices had been good lately, although the effects of the new component pricing formula remained an open question (Exhibit 3). The

Phoenix Farm's milk buyer had not yet introduced component pricing, but the family expected it soon.

The family obtained bids on rebuilding from different contractors, including the Amish laborers that Bill suggested. Still, the brothers felt they did not know much about what they could afford. Financial information about the business was held tightly by Luke. Although Emory had started using a computer program to track finances and generate a cash flow analysis back in 1992, he gave it up when Luke chose to ignore the computer analysis for tax preparation and business planning. Now Luke wasn't even saying what the insurance settlement was. Vern, Jr., had started keeping a new set of books using the CA Simply Money program since the first of the year, but his grandfather still hand-entered every transaction in a ledger. Emory observed that "Vern Jr. works for me but he knows the books better than I do." For his part, Vern, Jr., was not sure his computer efforts would prove any more useful than Emory's had.

Three days after the consultant's last visit, Bill phoned him and eagerly asked what the recommended parlor and herd size were going to be.

Study Questions

1. What is the decision?

2. Who is, or are, the decision maker(s)?

3. What alternatives do the partners have?

4. What information do the partners need to evaluate their alternatives?

5. What should the consultant recommend to the Phoenix Farm partners?

6. What should Vern, Jr. do?

Exhibits

1. Michigan Dairy Production Herd Evaluation Report for Phoenix Farm, September 30, 1994.

2. Excerpt from "Business Analysis for Specialized Michigan Dairy Farms: 1993 TelFarm Data."

3. Dairy outlook for 1994 from Michigan Farm News, Jan. 31, 1994.

EXHIBIT 1

Michigan Dairy Production Herd Evaluation Report

Phoenix Farm, September 30, 1994.

(Michigan Dairy Herd Improvement Association)

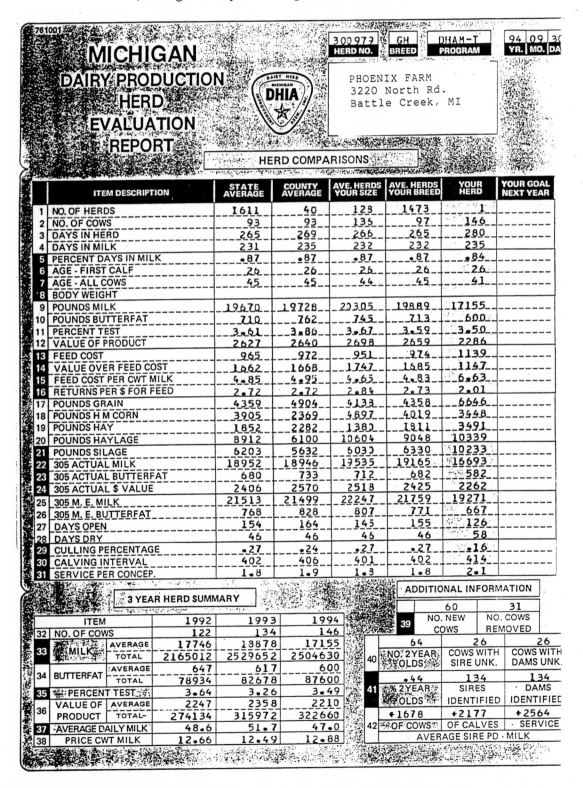

| | | HERD NO. 300979 | BREED GH | PROGRAM DHAM-T | YR. 94 | MO. 09 | DA 30 |

MICHIGAN DAIRY PRODUCTION HERD EVALUATION REPORT

DHIA

PHOENIX FARM
3220 North Rd.
Battle Creek, MI

HERD COMPARISONS

	ITEM DESCRIPTION	STATE AVERAGE	COUNTY AVERAGE	AVE. HERDS YOUR SIZE	AVE. HERDS YOUR BREED	YOUR HERD	YOUR GOAL NEXT YEAR
1	NO. OF HERDS	1611	40	129	1473	1	
2	NO. OF COWS	93	93	136	97	146	
3	DAYS IN HERD	265	269	266	265	280	
4	DAYS IN MILK	231	235	232	232	235	
5	PERCENT DAYS IN MILK	.87	.87	.87	.87	.84	
6	AGE - FIRST CALF	26	26	26	26	26	
7	AGE - ALL COWS	45	45	44	45	41	
8	BODY WEIGHT						
9	POUNDS MILK	19670	19728	20305	19889	17155	
10	POUNDS BUTTERFAT	710	762	745	713	600	
11	PERCENT TEST	3.61	3.86	3.67	3.59	3.50	
12	VALUE OF PRODUCT	2627	2640	2698	2659	2286	
13	FEED COST	965	972	951	974	1139	
14	VALUE OVER FEED COST	1662	1668	1747	1685	1147	
15	FEED COST PER CWT MILK	4.85	4.95	4.65	4.83	6.63	
16	RETURNS PER $ FOR FEED	2.72	2.72	2.84	2.73	2.01	
17	POUNDS GRAIN	4359	4904	4133	4358	6646	
18	POUNDS H M CORN	3905	2369	4897	4019	3448	
19	POUNDS HAY	1852	2282	1382	1811	3491	
20	POUNDS HAYLAGE	8912	6100	10604	9048	10339	
21	POUNDS SILAGE	6203	5632	6030	6330	10233	
22	305 ACTUAL MILK	18952	18946	19535	19165	16693	
23	305 ACTUAL BUTTERFAT	680	733	712	682	582	
24	305 ACTUAL $ VALUE	2406	2570	2518	2425	2262	
25	305 M. E. MILK	21513	21499	22247	21759	19271	
26	305 M. E. BUTTERFAT	768	828	807	771	667	
27	DAYS OPEN	154	164	145	155	126	
28	DAYS DRY	46	46	46	46	58	
29	CULLING PERCENTAGE	.27	.24	.27	.27	.16	
30	CALVING INTERVAL	402	406	401	402	414	
31	SERVICE PER CONCEP.	1.8	1.9	1.8	1.8	2.1	

3 YEAR HERD SUMMARY

	ITEM		1992	1993	1994
32	NO. OF COWS		122	134	146
33	MILK	AVERAGE	17746	18878	17155
		TOTAL	2165012	2529652	2504630
34	BUTTERFAT	AVERAGE	647	617	600
		TOTAL	78934	82678	87600
35	PERCENT TEST		3.64	3.26	3.49
36	VALUE OF PRODUCT	AVERAGE	2247	2358	2210
		TOTAL	274134	315972	322660
37	AVERAGE DAILY MILK		48.6	51.7	47.0
38	PRICE CWT MILK		12.66	12.49	12.88

ADDITIONAL INFORMATION

		60	31
39		NO. NEW COWS	NO. COWS REMOVED
40		64	26	26
		NO. 2 YEAR OLDS	COWS WITH SIRE UNK.	COWS WITH DAMS UNK.
41		.44	134	134
		% 2 YEAR OLDS	SIRES IDENTIFIED	DAMS IDENTIFIED
42		+1678	+2177	+2564
		OF COWS	OF CALVES	SERVICE
			AVERAGE SIRE PD - MILK	

EXHIBIT 2

Excerpt of pages 22-28 from

"Business Analysis Summary for Specialized Michigan Dairy Farms: 1993 TelFarm Data."

Agricultural Economics Report No. 578

Department of Agricultural Economics, Michigan State University

Table 29.

DAIRY FARMS WITH 150 OR MORE COWS

DATA FROM 49 SPECIALIZED MICHIGAN DAIRY FARMS, 1993

INCOME SUMMARY, AVERAGE CAPITAL AND ESTIMATED LABOR

		Per Tillable Acre		Per Milk Cow	
	Michigan Farm Total	Michigan Average	Your Farm	Michigan Average	Your Farm
1. Value of production	$ 594,951	$ 313	_____	$ 1,478	_____
Costs					
2. Labor	140,897	75	_____	345	_____
3. Power & equipment	124,948	111	_____	150	_____
4. Building & improvements	52,813	30	_____	124	_____
5. Crop supplies	58,516	72	_____	0	_____
6. Livestock services	114,567	0	_____	500	_____
7. Land charge	57,370	65	_____	19	_____
8. Other	21,718	5	_____	76	_____
9. TOTAL COST	$ 570,829	$ 358	_____	$ 1,214	_____
10. Management income a/	$ 24,122	$ -45	_____	$ 264	_____
11. Labor income b/	$ 39,781	$ -37	_____	$ 302	_____
12. Rate earned on owned capital c/ .	7.85%	-0.04%	_____	16.53%	_____
Farm Capital Owned					
13. Land (agricultural value)	$ 285,149	$ 558	_____	$ 9	_____
14. Buildings & improvements	161,524	62	_____	483	_____
15. Machinery	178,376	161	_____	202	_____
16. Livestock	398,423	0	_____	1,737	_____
17. Crops & supplies	165,716	203	_____	0	_____
18. Perennial crops	193	0	_____	0	_____
19. TOTAL OWNED	$1,189,381	$ 984	_____	$ 2,431	_____
Farm Capital Rented					
20. Land	$ 146,549	$ 472	_____	$ 0	_____
21. Other	35,202	23	_____	73	_____
26. TOTAL RENTED	$181,751	$ 495	_____	$ 73	_____
27. TOTAL CAPITAL (line 19 + 26) . .	$1,371,132	$ 1,479	_____	$ 2,504	_____
Labor					
33. Estimated labor hours	19,068	10.20	_____	46.74	_____

a/ Line 1 minus line 9.

b/ Line 10 plus line 651.

c/ Line 19 times .0581, plus line 10, divided by line 19, times 100.

Table 30.   FEED FED (FEED DISAPPEARANCE) IN DOLLARS, 150 OR MORE COWS a/

			Feed Per Milk Cow b/	
	Feeds	Michigan Average Total Farm	Michigan Average	Your Farm
60.	Corn	$ 47,739	$ 208	_____
61.	Corn silage	39,612	173	_____
62.	Oats	663	3	_____
63.	Barley	57	0	_____
64.	Hay equivalent	91,563	399	_____
66.	Pasture	1,281	6	_____
67.	Other feed costs	151,137	659	_____
68.	TOTAL FEED FED	$332,052	$1,448	_____

a/ Beginning inventory of feed crops and commercial feed plus feed crops produced plus feed and feed crops purchased minus sales minus the ending inventory of feed crops and commercial feed on hand at the end of the year.

b/ Includes feed fed to replacements and non-dairy livestock.

Table 31.   AVERAGE DAIRY FACTORS ON FARMS WITH 150 OR MORE COWS

		Total Farm		Per Milk Cow	
	Factor	Michigan Average	Your Farm	Michigan Average	Your Farm
80.	Number of cows (head)	229.40	_____		
81.	Milk sold (lbs.)	4,573,931	_____	19,939	_____
82.	Milk sales	593,327	_____	2,586	_____
83.	Cattle income	76,709	_____	334	_____
84.	Gross income dairy	670,036	_____	2,920	_____
85.	Return above feed	337,984	_____	1,472	_____
86.	Non-feed cost	278,384	_____	1,214	_____
87.	Return above all cost	59,600	_____	258	_____
88.	Calves born (head)	189.35	_____	0.83	_____
89.	Percent calf death loss	10.77	_____		
90.	Price/cwt. milk	12.97	_____		
91.	Net cost/cwt. milk	11.67	_____		

Table 32.   INCOME SUMMARY - VALUE OF FARM PRODUCTION, 150 OR MORE COWS c/

		Value of Farm Production			
	Income Source	Total Value		Value from Crops	Value from Livestock
		Average	Your Farm		
100.	Crops	$ 34,192	_____	$ 34,192	$ --
101.	Livestock	671,080	_____	--	671,080
102.	Feed Purchased	-151,138	_____	--	-151,138
103.	Internal feed crop transfer	--	_____	180,915	-180,915
104.	All other items	40,815	_____	40,815	--
105.	Total value of farm production (lines 100+101+102+103+104)	$594,949	_____	$255,922	$339,027

c/ Derived from the information on page 24 as follows: Sales minus purchases plus the change in inventory between the beginning and end of the year.

Table 33.   SALES AND PURCHASES OF INCOME ITEMS FOR VALUE OF PRODUCTION, 150 OR MORE COWS

	Dollar Sales		Dollar Purchases	
Item	Michigan Average (B)	Your Farm	Michigan Average (D)	Your Farm
500. Corn	$ 4,855		$ 7,723	
501. Corn silage	537		189	
502. Oats	99		40	
504. Barley	0		0	
505. Hay equivalent	669		3,541	
506. Pasture	0		241	
507. Wheat	2,799		0	
512. Dry edible beans	1,558		0	
509. Soybeans	5,173		0	
511. Other cash crops	1,380		0	
540. Crops Subtotal	$ 17,070		$ 11,734	
550. Dairy cattle	$ 70,492		$ 18,315	
570. Dairy products	593,327		0	
575. Other livestock items	1,215		277	
580. Livestock Subtotal	$665,034		$ 18,592	
600. Feed Purchased	$ 0		$151,336	
602. Forest products	$ 888		$ 0	
603. Custom work	2,183		0	
604. Refunds	5,630		0	
605. Government payments	27,560		0	
606. Other income	4,552		0	
607. Other Subtotal	$ 40,813		$ 0	

Table 34.   DOLLAR INVENTORY LEVELS FOR VALUE OF FARM PRODUCTION, 150 OR MORE COWS a/

	Beginning		Ending	
Item	Michigan Average (F)	Your Farm	Michigan Average (H)	Your Farm
500. Corn	$ 43,259		$ 59,846	
501. Corn silage	35,082		36,021	
502. Oats	395		464	
503. Growing alfalfa	4,661		4,419	
504. Barley	0		34	
505. Hay equivalent	41,189		49,737	
510. Growing wheat	745		950	
507. Wheat	352		257	
508. Dry edible beans	253		699	
509. Soybeans	3,747		6,012	
511. Other cash crops	2,719		2,819	
540. Crops Subtotal	$132,402		$161,258	
550. Dairy cattle	$385,492		$410,025	
575. Other livestock items	612		717	
580. Livestock Subtotal	$386,104		$410,742	
600. Feed Purchased	$ 4,559		$ 4,757	
602. Forest products	$ 0		$ 0	
603. Custom work	0		0	
604. Refunds	0		0	
605. Government payments	0		0	
606. Other income	10		12	
607. Other Subtotal	$ 10		$ 12	

a/ Machinery inventory changes are handled in the depreciation calculation. Changes in supplies on hand, like gasoline and fertilizer, are shown in Table 35.

Table 35.   FARM COSTS IN TOTAL AND ALLOCATED TO CROPS AND MILK COWS, 150 OR MORE COWS

Cost Items	Total Cash Costs		Total Noncash Costs & Inventory Changes	
	Michigan Average	Your Farm	Michigan Average	Your Farm
Labor (6.00 people, 18,011 hrs.)				
651. Operator, 2,409 hours at $6.50	$ 0		$ 15,659	
652. Family, 2,842 hours at $6.50	0		18,475	
653. Hired, 12,759 hours	106,763		0	
Machinery Expenses				
655. Repairs & vehicle maintenance	$ 46,356		$ -122	
656. Fuel, oil, and grease	14,535		75	
657. Custom hire and lease	11,608		-108	
658. Depreciation	0		42,106	
659. Interest allocation.a/	4,339		6,039	
Improvement Expenses				
661. Conservation	$ 385		$ 0	
662. Repairs	7,974		0	
663. Insurance	6,892		0	
664. Lease	6,719		0	
665. Depreciation	0		21,300	
666. Depreciation trees	0		5	
667. Interest allocation.a/	3,923		5,460	
668. Interest trees	5		7	
Crop Expenses				
670. Fertilizer and lime	$ 24,451		$ -839	
671. Supplies and packages	507		-395	
672. Seeds, plants + replacement trees	12,068		-219	
673. Chemicals	11,318		855	
674. Marketing	60		0	
675. Other + dryer and irrigation fuel	1,154		0	
676. Interest allocation.a/	4,031		5,610	
Livestock Expenses				
678. Semen + breeding supplies	$ 6,677		$ -228	
679. Veterinary, medicine and drugs	20,727		0	
680. Marketing, trucking, commercial, etc . . .	36,465		0	
681. Livestock supplies	13,357		0	
682. DHIA, registration, bedding + other	14,430		0	
683. Interest allocation.a/	9,691		13,488	
Land Charge (Expense)				
685. Tax	$ 16,730		$ -356	
686. Interest allocation on land.a/	6,936		9,653	
687. Rent	24,559		0	
Other Expenses				
689. Utilities	$ 15,570		$ 0	
690. Miscellaneous	6,148		0	
691. TOTAL	$434,378		$136,465	

a/ Interest is computed by multiplying the values in Table 29 above line 19, by the 5.81 percent rate. Interest listed as cash cost is interest reported as paid during the year.

Table 35. (continued)   DAIRY FARMS 150 OR MORE COWS

	Costs Allocated to Crops				Costs Allocated to Milk Cows		
Cash	Non-Cash & Inventory Change	Per Tillable Acre — Michigan Average	Per Tillable Acre — Your Farm	Cash	Non-Cash & Inventory Change	Per Milk Cow — Michigan Average	Per Milk Cow — Your Farm
$ 0	$ 6,854	$ 8.38		$ 0	$ 8,805	$ 38.39	
0	8,086	9.89		0	10,389	45.29	
46,728	0	57.13		60,034	0	261.70	
		$ 75.40				$ 345.38	
$ 27,341	$ -72	$ 33.34		$ 19,015	$ -50	$ 82.67	
12,567	65	15.44		1,967	10	8.62	
11,608	-108	14.06		0	0	0.00	
0	31,371	38.35		0	10,735	46.80	
3,211	4,468	9.39		1,128	1,570	11.76	
		$ 110.58				$ 149.85	
$ 385	$ 0	$ 0.47		$ 0	$ 0	$ 0.00	
4,703	0	5.75		3,271	0	14.26	
2,656	0	3.25		4,236	0	18.46	
6,719	0	8.21		0	0	0.00	
0	6,829	8.35		0	14,470	63.08	
0	5	0.01		0	0	0.00	
1,232	1,714	3.60		2,692	3,746	28.06	
5	7	0.01		0	0	0.00	
		$ 29.65				$ 123.86	
$ 24,451	$ -839	$ 28.87		$ 0	$ 0	$ 0	
507	-395	0.14		0	0	0	
12,068	-219	14.49		0	0	0	
11,318	855	14.88		0	0	0	
60	0	0.07		0	0	0	
1,154	0	1.41		0	0	0	
4,031	5,610	11.79		0	0	0	
		$ 71.65				$ 0	
$ 0	$ 0	$ 0		$ 6,677	$ -228	$ 28.11	
0	0	0		20,727	0	90.36	
0	0	0		36,465	0	158.96	
0	0	0		13,357	0	58.23	
0	0	0		14,430	0	62.91	
0	0	0		9,691	13,488	101.04	
		$ 0				$ 499.61	
$ 12,505	$ -266	$ 14.96		$ 4,225	$ -90	$ 18.02	
6,885	9,583	20.13		51	71	0.53	
24,559	0	30.03		0	0	0.00	
		$ 65.12				$ 18.55	
$ 1,526	$ 0	$ 1.87		$ 14,044	$ 0	$ 61.22	
2,691	0	3.29		3,457	0	15.07	
		$ 5.16				$ 76.29	
$218,910	$73,548	$357.56		$215,467	$62,916	$1,213.54	

Table 36.

CROP YIELDS PER ACRE ON FARMS WITH 150 OR MORE COWS

	Owned Land		Rented Land	
Item	Michigan Average	Your Farm	Michigan Average	Your Farm
851. Corn grain, bushel	118.6 _____		106.6 _____	
852. Corn silage, ton	16.4 _____		14.0 _____	
853. Oats, bushel	49.0 _____		52.6 _____	
855. Hay equivalent, ton	4.6 _____		3.7 _____	
858. Pasture, ton	2.5 _____		2.2 _____	
860. Wheat, bushel	48.6 _____		38.8 _____	
864. Soybeans, bushel	41.7 _____		40.2 _____	

Table 37.

AVERAGES OF CROP ACRES GROWN AND VALUES PER ACRE, 150 OR MORE COWS

	Acres		Crop Value Per Tillable Acre a/	
Item	Owned	Rented	Owned	Rented
851. Corn grain	139.0	95.7	273	$246
852. Corn silage	81.4	45.8	339	290
853. Oats	5.4	4.6	77	83
854. Barley	0.0	1.0	N/A	90
855. Hay equivalent	175.7	133.9	341	279
856. Rye	0.9	1.6	55	77
858. Pasture	11.2	7.3	59	53
859. Wheat	12.8	6.7	158	126
864. Soybeans	26.6	9.1	267	257
865. Other cash crops	14.6	5.1		
873. Idle tillable	39.7	0.0		
874. Total tillable	507.3	310.8		
875. Woodland	78.3	0.0		
876. Non-tillable	15.2	0.0		
877. Total acres	600.8	310.8		

a/ Computed by multiplying yield per acre times average of prices used on individual farm reports.

Table 38. COMPARATIVE INCOME STATEMENTS OVER FOUR YEARS

AVERAGE OF ALL SPECIALIZED DAIRY FARMS

	1990	1991	1992	1993
Revenue:				
Milk	$273,915	$238,622	$285,693	$311,429
Dairy Cattle	35,921	37,292	38,210	39,584
Crop Sales	15,201	19,056	15,819	17,375
Refunds	4,542	4,054	4,332	4,449
Government	7,104	7,319	8,713	18,103
Other Income	2,360	2,797	2,451	4,405
Total Revenue	$339,043	$309,140	$355,218	$395,345
Cost of Goods Sold: (1)				
Inventory Change (+ or -) . . .	$+ 9,986	$ +7,605	$ +8,210	$ +26,796
Less Purchased	- 4,993	-4,462	-9,897	-9,438
Net Change in Revenue	$ +4,993	+3,143	-1,687	+17,358
Gross Farm Profit	$344,036	$312,283	$353,531	$412,703
Operating Expenses:				
Hired Labor	$ 33,846	$35,913	$41,626	$49,156
Repairs	26,652	24,082	27,001	30,275
Fuel & Oil	8,836	7,857	7,945	8,838
Custom Hire	5,390	5,226	6,681	6,851
Land Rent	9,870	11,622	12,312	15,214
Insurance	3,669	3,967	4,057	4,516
Fertilizer	14,637	12,650	13,899	15,201
Seed	6,661	6,028	7,517	7,766
Chemicals	5,515	5,711	6,204	7,089
Conservation	260	137	199	175
Crop Supplies	607	392	391	455
Crop Marketing	101	157	159	145
Other Crop Expense	771	539	665	1,263
Feed	58,479	55,027	68,169	81,768
Semen & Breeding	3,334	2,856	3,442	3,445
Veterinary & Medicine	6,776	7,053	8,293	9,858
Livestock Marketing	14,123	14,575	17,541	19,825
Livestock Supplies	10,250	10,771	12,437	13,959
Taxes	10,252	9,702	8,949	10,286
Utilities	7,468	7,291	7,885	8,954
Interest	17,966	16,965	14,839	15,443
Miscellaneous	2,964	2,827	3,258	3,858
Depreciation	29,773	26,524	29,989	36,117
Total Operating Expense . . .	$278,200	$267,872	$303,458	$350,457
Change Expense Accrued				
Increase (-), Decrease (+) .	+ 9,452	+1,463	+2,501	+2,073
Adjusted Total Operating Expense	$287,652	$269,335	$305,959	$352,530
NET FARM INCOME FROM OPERATIONS	$ 56,384	$ 42,948	$ 47,572	$60,173

Exhibit 2, continued

DEFINITION OF TERMS USED IN THIS REPORT
(Does not include landlord's income or costs)

GENERAL FARM BUSINESS FACTORS

Tillable acre: This includes the land in harvested crops, land devoted to crops that failed, tillable pasture, tillable land reserved for government programs, and idle tillable land.

Person equivalent: The number of hours of hired labor, operator's labor, and family labor divided by 3,000.

Total investment: This is the average inventory value of land, buildings (less the farm dwelling), machinery, livestock and feed. The inventory value of buildings and machinery is based on cost less depreciation claimed for income tax purposes. Bare land is assigned a conservative market value based on quality.

Crop value: Crop value is calculated by multiplying the yield of each crop by a price suggested by the farm owner when completing a crop inventory or by the following prices:

Corn open market	2.55 per bu.	Wheat	$ 3.25	per bu.
Oats	1.45 per bu.	Alfalfa hay med qual.	70.00	per ton
Navy Beans	17.50 per cwt.	Soybeans	6.40	per bu.
Dark red beans	30.00 per cwt.	Growing wheat	60.00	per acre
Light red & cranberry beans	24.50 per cwt.	Corn silage untreated	19.00	per ton

The crop value for other crops is calculated by subtracting the total of purchases and the beginning inventory from the total of cash sales and ending inventory.

Crop Income: This is the sale of the crop plus the change in inventory.

Value of farm production: This is cash sales in the farm business minus the purchases of feed, feed crops, and livestock plus the change in inventory of crops and livestock.

COST FACTORS
(Does not include landlord's costs)

Total cost: This is the sum of items listed below. It includes cash expenses and changes in inventory for supply items that are inventories such as power, fuel, fertilizer, pesticides, etc. It also includes noncash costs for operator's and unpaid family labor, depreciation on improvements and machinery, and interest on the owner's equity.

Labor: This includes the cash cost for hired labor including paid perquisites and social security plus the value of operator's and unpaid family labor at $6.50 per hour.

Power and machinery: This includes the following items: repairs and supplies for upkeep on machinery including tractors; repairs and upkeep on trucks and farm share of automobiles including fuel for these if so reported; gas, oil, and grease for tractors, custom hire of machinery, depreciation on machinery, and interest 1/ on investment in machinery.

Improvements: This includes repairs on buildings, fences and wells and bulldozing, cleaning ditches, fence rows, etc., which are classified as conservation expense for income tax purposes. It includes fire and wind insurance premiums, depreciation, and interest on improvement investment. Interest on trees is primarily for fruit orchards.

Crop expense: This includes fertilizer, lime, seed and plants, spray materials and herbicides, crop insurance, crop marketing, interest on the crop inventory, and other crop expenses.

Livestock expense: This includes breeding, veterinary and medicine, milk and livestock marketing, milkhouse supplies, registration, advertising, heat for livestock buildings, and other livestock service and supply items.

Land charge: This is interest for the investment in land plus taxes paid. Land values are estimated by Extension staff members. An effort is made to keep them comparable between farms. An attempt is made to use an agricultural value and not reflect urban real estate values.

Other expenses: These include utilities and other miscellaneous expenses not included elsewhere.

EARNING FACTORS
(Does not include landlord's earnings)

Management income: Value of farm production less total costs.

Labor income: Management income plus the value of operator's labor.

Rate earned on investment: Management income plus interest on investment divided by the total operator's investment times 100.

1/ Interest on total investment is reported in two segments -- interest paid and interest on equity. Interest paid is reported by Telfarmers. The amount of debt is estimated by capitalizing the amount of interest reported at 7.5 percent. The estimated debt is subtracted from the total investment to determine equity. Interest is computed on the equity at 5.0 percent. Interest is allocated to machinery, improvements, land, livestock, and crop and orchard inventory according to investment.

EXHIBIT 3

Dairy Outlook for 1994 from Michigan Farm News, Jan. 31, 1994.

Michigan Farm News

12 *Dairy Outlook - Purchased Feed Costs up 10–20 Percent*

Higher feed costs will require strict attention to management details in 1994.

Januay 31, 1994

Larry G. Hamm and Sherrill B. Nott, Dept. of Agricultural Economics, MSU

National

U.S. milk production will be around 152 billion pounds for 1993, about the same as it was in 1992. But, there were notable shifts within the country. Total production was down in Minnesota and Wisconsin, but was up in California and the West.

Expect the financial strains caused by flooding and poor quality feed to plague the upper Midwest through the first half of 1994. Other regions, though, will easily offset any local production drops. The country may see 1994 production 1 percent higher than in 1993.

The U.S. dairy herd will likely decline 1.3 percent, ending 1994 at 9.6 million head. Production per cow should increase more than 1.6 percent and will likely approach 16,000 pounds per cow. Credit the exit of many lower producing cows in the upper Midwest and the use of bovine somatotropin (BST) for the higher average milk per cow.

Commercial disappearance was around 144 billion pounds in 1993, up nearly 1.6 percent from 1992. Expect a 1 percent increase in 1994 with demand reaching 145.5 billion pounds. Commercial disappearance will follow population growth and moderate increases in personal income.

In early December, the USDA predicted support purchases of both butter and nonfat dry milk were likely after holiday needs had been met. For all of 1993, removals of milkfat were projected to decrease 2-3 billion pounds, milk equivalent, from 1992's 10 billion. Skim solids removals were expected to rise to about 3 billion pounds, milk equivalent, from 2 billion a year ago.

USDA projected that 1994 surplus milk production would be between 6 and 7 billion pounds milk equivalent, regardless of measure. The surplus of skim solids may exceed the milkfat surplus for the first time since 1987.

This level of surplus means that producer assessments under the Omnibus Budget Reconciliation Act of 1990 will likely increase to more than 16 cents for May

through December. The check-off deduction for advertising and promotion by the National Dairy Board will continue.

The national average price for milk at the farm was $12.80 in 1993, down about 3 percent from 1992. With the developing surplus situation predicted by the USDA, the national average milk price could drop to $12.00 in 1994. Given the uncertainty of major factors in the supply and demand picture in the next few months, this price estimate could be off by a dollar, plus or minus.

BST

Our country's Food and Drug Administration in early December approved BST for farm use. With the 90-day delay tacked on by Congress, it may be available by early February. However, on Dec. 24, 1993, the *New York Times* reported that the Foundation un Economic Trends would file a lawsuit in January attempting to block BST sales. We'll soon know!

If such roadblocks do not occur, it will still take awhile for this new technology to be adopted and have an impact. Michigan will probably adopt its use as quickly as anywhere in the country.

Michigan Pricing

The Minnesota Wisconsin (M-W) price series is a basic mover of milk prices in Michigan. And, it will continue to be as we go to component pricing. The month-to-month changes in the M-W averages were close to 50 cents per cwt. for 1993. The industry had a double peak in the M-W in 1993; it's questionable whether that will happen again, but expect monthly price volatility during 1994.

The drastic cutbacks in total milk supplies in Minnesota and Wisconsin during the final months of 1993, compared to the previous year, have tended to offset much of the expected seasonal drop in the M-W series. This will support the Michigan farm price for the first quarter of 1994.

Multiple component pricing will bring a new way to calculate individual farm milk prices in Federal Order 40. In addition to the M-W base and butterfat content, the price will also be influenced by the percent of protein in the milk and the level of somatic cell counts. This change will likely happen sometime in 1994.

Those selling into the Ohio market are already under a component pricing system. The Upper Peninsula, under Order 44, will not change to component pricing this year.

Farm Costs

The USDA tracks the prices paid for 15 farm production items. In 1977, the index for these items was set to 100. By October 1993, the grouping that had increased in price the most was automobiles and trucks. Its index was 276.

Other farm machinery was second with 248 and tractors and self-propelled equipment was third with 237. Labor was fourth at 221. There is reason to believe that these groupings will continue to increase in cost faster than the others.

The categories that had increased the least by October 1993 were feed and fertilizer. They tied with an index of 127 each. Expect fertilizer prices to continue low to lower as crop farmers attempt to more closely match plant needs with nutrients available resulting in less fertilizer purchases.

Feed Costs: Feed prices will be higher in 1994, especially the first half of the year. Excess rain lowered forage quality in the upper Midwest. Flooding took out a lot of corn and soybeans. By the end of 1993, the cash prices of corn and soybean meal were rising rapidly.

When 1994 is over, it's estimated that the per cow purchased feed costs will be 10-20 percent higher than in 1993. That's assuming normal weather for the 1994 crop year and the chance to rebuild feed inventories at closer to long-run average prices.

The impact of feed prices will vary from farm to farm within the state. In southern Michigan where corn often yielded well, dairy farms will have enough to feed and excess to sell as a cash crop. Such farms may be better off on a cash flow basis than they were in 1993. However, purchased protein feeds will be a problem.

Fuel Costs: Fuel prices were headed lower in th closing weeks of 1993 as the OPEC cartel was unabl to stop crude oil prices from dropping. Although reta gasoline prices were down 10 percent in sout Michigan, it is not known how long the situation wi last. If excess crude stays on the market for severa months, it could noticeably lower the fuel costs planting and ease the pressure on trucking costs for mil and a variety of other farm products.

Taxes: Real estate tax levels are an unknown as this i written! The Michigan Legislature has presented a pla for restructuring how schools will be financed. Voter will decide upon the final version. It appears land taxe will be reduced. The cash flow impact after the substitu tion of different taxes is not clear.

The effect on income taxes and the moderating adjust ments for those with P.A. 116 contracts is current unknown. On a pre-income tax basis, the proposed lan tax changes could mean more farm profit.

Overall Expenses: A sample of 200 farms in 199: averaging 113 cows had total operating expense o $303,458 including depreciation and cash interest, bu not including unpaid labor. Cash feed purchases wer 22 percent, land taxes were 2.9 percent and fuel cost were 2.6 percent of that total. The expected increase in purchased feed costs may well more than offset any decreases in land taxes, diesel fuel and gasoline. A dair farm with the same size and production levels as in 1993 can expect higher total costs in 1994.

Management Critical

Given the above, managers will be looking to improve efficiency. New technology such as BST will be attractive. Managers will be looking to improve their skills in seminars such as those in the Animal Management Advancement Project.

The long-run weather forecast is for a warmer and wetter than usual situation from now until the start of the growing season. This has implications for selecting next year's crop varieties. If stored forages are in short supply, planning on how many head to keep will pay dividends. Perhaps this will be the spring to try grazing those dry cows and heifers, if not the milkers. They can get onto moist ground and harvest the first plant growth far earlier than machinery can.

The information age is upon us, and it needs to be managed. Computers can help. They must not be part of the machinery cost indexes mentioned above, because this year's computers are more powerful and sell for about half what they did three years ago.

Consider the data and information a Michigan dairy farmer could handle in 1994. The DHIA data will be input at the farm on a portable computer; a report can be left behind for immediate managerial action.

As BST use is started, individual cow performance will need to be tracked to see if the response is profitable. Multiple component pricing puts a premium on tracking milk protein levels by cow as rations change.

Plant nutrient management is increasingly critical as environmental regulations pertaining to record keeping come on line. Electronic message sending systems are the fastest way to tap expertise around the world, as well as across the state. The outlook is that managers will turn to 486 computer systems for help in all of these chores.

Is a Drink of Water a Drink of Weed Killer?

By Susan Shumaker, Marla Reicks and Roger Becker, University of Minnesota

Phil Peterson, the utilities superintendent in Oakmont, put down Wednesday's newspaper (Exhibit 1). He pushed his chair back from his desk and thought about the article he had just read. "Do people think a drink of water is a drink of weed killer?" he wondered. The article covered the results of a study that the Environmental Working Group (EWG) had released to the media yesterday. The EWG is a consumer interest group that conducts research about environmental issues that affect the public.

Farm Chemicals in the Drinking Water?

The study results released by the EWG claimed that farm chemicals (herbicides, or weed killers) have seeped into the drinking water in much of the Midwest, exposing 14 million Americans to an increased risk of cancer. In Minnesota, those at greatest risk lived in Oakmont where lifetime cancer risks were estimated by the EWG to be 30 times higher than the standard. The article in today's paper gave highlights of the study and reactions from the federal EPA, herbicide manufacturers, and the American Crop Protection Association (an association of chemical manufacturers).

Jay Vroom, the president of the American Crop Protection Association, said that if the study results were taken out of context, it would be like pouring gasoline on the fire of public opinion. Phil knew exactly what Jay meant. The quote from the EPA administrator suggested that the study was a cause for concern, but not alarm.

As the utilities superintendent, Phil was accountable to Oakmont residents for their water quality. He routinely fielded questions when issues arose concerning water quality in Oakmont. Phil knew that it would not be long before people in Oakmont would call to find out what he thought of the

results of the study. They would be afraid, angry, upset, and want answers.

Risk Communication

Phil knew that some people would not listen to his explanations or believe what he had to say. Risk communication is critical in situations like this one. He thought again about the report. The EWG believes that herbicides used by Minnesota farmers are ending up in the drinking water at levels that expose Oakmont infants to 90 percent of the "acceptable" lifetime exposure to the chemicals by their first birthday. By the time the child is six, he or she has exceeded the lifetime cancer risk standard by a factor of four. That was scary stuff!

The data used by the EWG was obtained from the U.S. Geological Survey several years ago. Phil questioned why the EWG didn't gather its own data. EWG claimed that the two main drinking water sources for Minnesota are contaminated with herbicides. They maintained that the drinking water tests in Minnesota are not specific enough and do not test for the most toxic herbicide, cyanazine. Also, EWG alleged that the tap water contained five common weed killers that are not effectively removed with current water treatment methods. The EWG wanted tougher federal standards.

Recent Tests Showed Oakmont's Water Safe

Just over two months ago, the Department of Health tested the Oakmont water supply. It fell within the standards of the Safe Drinking Water Act. Phil decided to call the Minnesota Department of Health and talk with someone who tests municipal water supplies.

Phil talked to Bob Carolan, a spokesman for the Department of Health. Phil was reassured by Bob that Oakmont's water supply was tested properly and found to be safe. Bob said, "As far as the Department of Health is concerned, there isn't any hazard in drinking treated water from Oakmont's plants." He also said that the EWG tested water that had not been treated. The EWG assumed that because they detected herbicides in untreated water, they would also find herbicides in treated water. Bob told Phil that the EWG did a great job misrepresenting the facts. He told Phil about a report that he had seen earlier that day responding to EWG's claims. It explained in detail how the EWG misrepresented the facts (Exhibit 2). Bob promised to fax a copy of the response to Phil right away.

Phil put down the phone and thought about the people living in Oakmont. The claims made by the EWG were alarming, and Phil knew that the people in Oakmont would be worried. He needed to respond quickly and knew that gaining the confidence of the people would not be easy. They had heard too many times about things they thought were safe and then later found were not. Once people were fearful about the safety of the water supply, they might be suspicious or skeptical of any information that Phil might share with them that tried to disprove the EWG's claims.

How Much Information to Provide?

Under these circumstances, Phil felt that it would not be a good idea to give out a mass of technical information that could be confusing.

People would have questions about the level of risk involved with herbicides. Like most things, herbicides included, "zero risk" is unrealistic. So how do you address the issue of acceptable risk when some people believe that there is no acceptable level of exposure to cancer-causing chemicals? The current standards are based on "no significant risk" levels where exposure could result in one additional cancer in one million people. Risk assessment was a complicated concept to explain in a short phone conversation.

Phil picked up a fax that his secretary had put on his desk (Exhibit 2). He scanned the information and thought some more about how he would communicate all of the risk issues involved. His credibility was important. He knew he had to come up with a reasonable, sound explanation for what people were reading in this morning's paper. People would want to know what he was going to do about the results of the report. His secretary interrupted his thoughts and told him that a worried mother of a six-month-old baby was on the line. What should he tell her?

EXHIBIT 1

12 - Wednesday, Oct. 19, 1994 The Free Press, Oakmont

Group: Weedkillers foul Oakmont water

EPA says herbicide report cause for concern, not alarm

Free Press Washington Bureau and The Associated Press

WASHINGTON -
Oakmont's drinking water contains dangerously high levels of weedkillers, an environmental group contended Tuesday.

The Environmental Working Group says herbicides in Oakmont drinking water are linked to cancer rates that are 30 times the accepted risk level recognized by the federal Environmental Protection Agency.

The Oakmont claim was part of a general report issued by the group, which says the tap water of millions of Midwesterners is contaminated by five common herbicides.

Cyanazine is the primary contaminant affecting the Oakmont area, the group said. The group's

conclusions are drawn from a study conducted during 1990-1991 by the U.S. Geological Survey.

The environmentalist group claims that infants in Oakmont receive 90 percent of the acceptable "lifetime" exposure to the chemicals by the age of 1.

But the group made its own determination of what constitutes safe, arguing that federal standards lag behind. The group used the most conservative possible benchmark of cancer risk in arguing that some supplies were dangerously contaminated.

Phil Peterson, Oakmont's utility superintendent, was unavailable for comment Tuesday on the study.

Between 150,000 and 300,000 pounds of the five chemicals are applied yearly in White Water and neighboring counties.

Cyanazine is sold by DuPont as Bladex, a weedkiller popular among the state's corn farmers. The group claims the chemical causes birth defects and cancer, but the EPA Office of Ground Water and Drinking Water does not regulate cyanazine as a carcinogen.

A 1993 study of Oakmont water by city officials found no pesticides, the environmentalists noted. But the group said that survey was flawed because Oakmont's public works department does not monitor for the presence of cyanazine.

The federal Environmental Protection Agency said the report should be viewed with concern but not alarm.

"This study is another in a series of wake-up calls that tells us we can no longer take for granted that our drinking water is safe all the time, "

said EPA Administrator Carol M. Browner.

The pesticide industry questioned the methods used in the study, which was released to the media before any outside scientists or industry groups could review it.

"They're trying to manufacture fear out of a situation that has a lot of data which is easily taken out of context and poured like gasoline on the fire of public opinion," said Jay J. Vroom, president of the American Crop Protection Association.

Vroom said each compound undergoes more than 120 different health and environmental studies, that includes laboratory testing on immature and mature animals, including pregnant ones. The EPA routinely asks for new data.

EXHIBIT 2

Environmental/Food Safety Decision Case Series

Faxed Report from the Department of Health

Tap Water Blues or Blue Ribbon Tap Water?

The report, *Tap Water Blues*, prepared by the Environmental Working Group and released through a series of news conferences on October 18, 1994, presents the most convincing evidence yet assembled that the occurrence of herbicides in Midwestern drinking water supplies does not pose a significant cancer risk. Unfortunately, the Environmental Working Group misrepresented the results of their own study, and, instead used it to raise unwarranted fears regarding the safety of Midwestern drinking water.

The new analyses conducted by the Environmental Working Group set out to answer a very appropriate question--What cancer risks are associated with existing concentrations of the five major herbicides in Midwestern drinking water? Only one of these five herbicides (alachlor) is currently regulated by the EPA on the basis of its cancer causing activity. For the other four herbicides (atrazine, cyanazine, simazine, and metolachlor) the EPA considers other toxicity characteristics to be more important than their cancer causing properties and therefore regulates them on the basis of the other toxic characteristics.

In no state are herbicides responsible for even one additional cancer case per year in the assessed population. The Environmental Working Group data indicates that among the 12,000,000 Midwestern residents having the highest levels of herbicide contamination, herbicides would be responsible for no more than three cancer occurrences per year.

Cancer is a major cause of death and disease in the United States. What proportion of these are due to herbicides? The death rate from cancer in the United States is about two thousand per million per year. However, no information is available regarding the fatality rate for herbicide induced cancers, so the percentage of cancer deaths attributable to herbicides cannot be calculated with any confidence.

Perhaps it is not surprising that in *Tap Water Blues*, the Environmental Working Group is careful not to make any estimates of the actual number of cancers that may result from herbicides in drinking water. Rather, they restrict their comments to risk levels relative to benchmark risks of 1 in a million. They clearly state that these risks are totally unacceptable and that herbicides represent very significant human health threats in Midwestern drinking water. They warn us that this unsafe water containing unacceptable cancer-causing herbicides is being used by 65,000 infants in the Midwest. Through this approach, they effectively strike fear into the hearts of millions of Midwestern residents. To protect Midwesterners, they call for a complete ban of the use of three of the offending herbicides.

What kind of attention would the Environmental Working Group have obtained if they had reported that herbicides were responsible for three additional cancer cases per year among the 12,000,000 Midwestern residents whose water has the highest herbicide concentrations? How many headlines would that story have attracted? Could they justify calling for a complete ban of the use of the offending herbicides within a two year period?

If the Environmental Working Group is really interested in reducing the cancer risks in the United States, why do they focus on herbicides which their own assessments suggest account for no more than 0.006% of cancer cases? Certainly, we should take effective and efficient steps to reduce cancer occurrences throughout the world, but focusing on these herbicides is not an efficient way to do so. Given the minimal cancer occurrences attributable to herbicides in Midwestern water supplies, perhaps Midwestern farmers and the agrichemical industry should be given a blue ribbon for developing food production technologies that result in such small cancer risks in our drinking water supplies.

David B. Baker and R. Peter Richards, Water Quality Laboratory, Heidelberg College, Tiffin, Ohio 44883, 10/20/94

Black-Mason Farm: "What Now?"

**By A.L. Skidmore, J.W. Lloyd, and B. Dartt,
Michigan State University**

The long string of Dairy Herd Improvement Association (DHIA) awards hanging on the kitchen wall are only a small indicator of the accomplishments of the Black-Mason Farm. The management team consists of Elmer Black and Gerry Mason. Their wives also play integral parts in the operation, helping with long-term strategic planning, daily chores and bookkeeping. Elmer's wife, Marie, is Gerry's sister. Marie and Gerry's father began the dairy in the mid-1960s', but the farm has been in the Mason family for five generations. The farm is owned and operated as a partnership formed by Elmer and Gerry's families. They milk 187 Holstein cows and farm about 2,400 acres of soybeans, dry beans, wheat, alfalfa, and corn.

The Management Team

Management of this operation is divided between the brothers-in-law. Elmer has responsibilities similar to a general manager. In addition, he is primarily responsible for the cow herd. Gerry is primarily responsible for the replacement herd, the equipment, and the field work. Both Elmer and Gerry milk the cows with help from other full-time and part-time employees. Many of the employees are also family members. Elmer constantly seeks the advice of others to help improve management of the farm. Gerry is not as proactive as Elmer in seeking advice from others.

The Last Big Decision

In 1990, the farm was at a major crossroads. A decision had to be made to stay in dairy farming or get out. The herd was producing very well (Exhibit 1) but somatic cell count levels were more than 500,000 cells per ml. of milk. The national legal limit had just been lowered to 750,000 cells per ml. of milk. The contagious mastitis organism,

Staphylococcus aureus, had been cultured from over 75 percent of the cows.

The lactating cows were housed and managed as one group. There was no way to group the cows in the existing facilities. To manage the mastitis problem, the cows would have to be grouped. This would require a new barn. This forced the issue: either get out of dairy farming and concentrate on cash crops; or take the plunge, go into debt, and build a new barn. The decision was made to build a new barn and take advantage of current management and technological advancements in the dairy business. The new 188-freestall lactating cow barn was built in 1991.

The Dairy Enterprise

Elmer and Gerry have worked hard and are proud of the level of production they have achieved. By 1993, their rolling herd average milk production is 25,378 lb. milk per cow per year, as calculated by Michigan Dairy Herd Improvement Association. This is well above the state DHIA average of 19,262 per lb. (Exhibit 1). Somatic cell count is about 200,000 cells per ml. The mastitis situation described earlier is under control, but not eliminated. Average age at first calving is about 32 months. The herd is milked three times per day in a double-7 herringbone parlor. A nutritionist balances the cow herd rations regularly. All animals are bred by artificial insemination.

The lactating cows are separated into three groups to control mastitis. These are: a *Staphylococcus aureus* positive group, a *Staphylococcus aureus* negative group, and a group of recently-calved cows for which culture results have not yet been received.

Lactating cows are housed in a freestall barn with curtain sidewalls and natural ventilation. The walk alleys are cleaned automatically with alley scrapers. The dry cows and calves, up to about four-months-old, are housed in the old freestall barn on the opposite side of the parlor from the new barn. Newborn calves are individually grouped and housed until they are weaned. After weaning, they are moved into transition groups of about 10 to 15 calves. The transition pen is at one end of the dry cow barn. After they graduate from the transition pen, they are moved to the farm down-the-road. Heifers are grouped by age into three groups (Exhibit 2). The younger calves are on the bedded packs. Heifers are housed and managed as one group, from about seven-months-old until six weeks before calving. They are then moved to the main farm with the dry cows. The heifer facilities, a quarter-mile away, are very old, inefficient, and in great need of repair. There is adequate space for the heifers and the pens are cleaned regularly to keep the heifers clean and dry.

All the heifers receive a diet of minerals and a 50:50 mixture of corn silage and alfalfa haylage. The calves on the bedded pack also get additional grain placed on top of the forage once a day.

The Current Situation

While providing advice for the mastitis problem, many people have suggested that something needs to be done about the age of heifers at first calving. The current average age at first calving is 32 months, compared to the state average of 26 months (Exhibit 1). Current agricultural Extension recommendations suggest calving at 24 months. Elmer is reminded of this every month when one of the advisors visits the farm. Elmer's response, "If I tell Gerry to better manage the heifers, he'll just tell me to do it myself, and I don't have the time to manage them and everything else I do on the farm. Why don't you talk to Gerry about the changes that need to be made." This same conversation occurs many times, but nothing ever happens.

In the spring of 1993, all the heifers at the farm down-the-road were weighed. Birth and breeding dates were recorded. Exhibit 3 is a graph of body weight and standard growth lines. The solid line in Exhibit 3 represents the growth of a calf reaching 1,250 pounds by 24 months. The broken line indicates the growth of a calf reaching 1,250 pounds by 28 months. Exhibit 4 shows the age distributions of bred heifers at first breeding, and heifers more than 10-months-old, but never bred. The current average age at first breeding is 16.9 months. Fifty-two percent of all heifers more than 15 months of age have never been bred.

In October 1993, Elmer decided to calculate what it costs to calve heifers at 32 months instead of 24 months. The results of his calculations are shown in Exhibit 5.

After seeing what the extra eight months are costing them, Elmer and Gerry got excited about doing something different. "At that cost," said Elmer, "a new heifer barn would easily pay for itself."

Study Questions

1. What is the problem?
2. What is the cause of the problem?
3. Will a new barn solve the problem?
4. What are alternative solutions to the problem?
5. Do you agree with the partial budget in Exhibit 5?

EXHIBIT 1

Black-Mason Farm and state DHIA averages.

Year	Black-Mason Farm DHIA rolling herd average	DHIA state average rolling herd average	DHIA state average age at first calving	Number of cows on Black-Mason Farm
1983	20,336	15,984	27	112
1984	20,108	16,107	27	104
1985	21,562	16,534	27	90
1986	21,919	17,041	27	111
1987	21,381	17,083	27	119
1988	20,284	17,464	27	112
1989	19,567	17,837	27	104
1990	21,474	18,153	26	121
1991	21,320	18,876	26	144
1992	25,283	19,262	26	167
1993	25,378	19,839	26	187

EXHIBIT 2

Current Heifer Raising Facilities

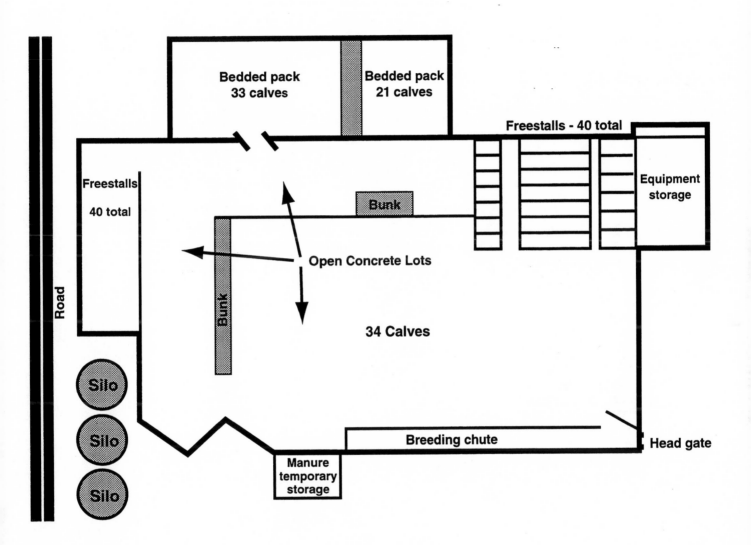

EXHIBIT 3

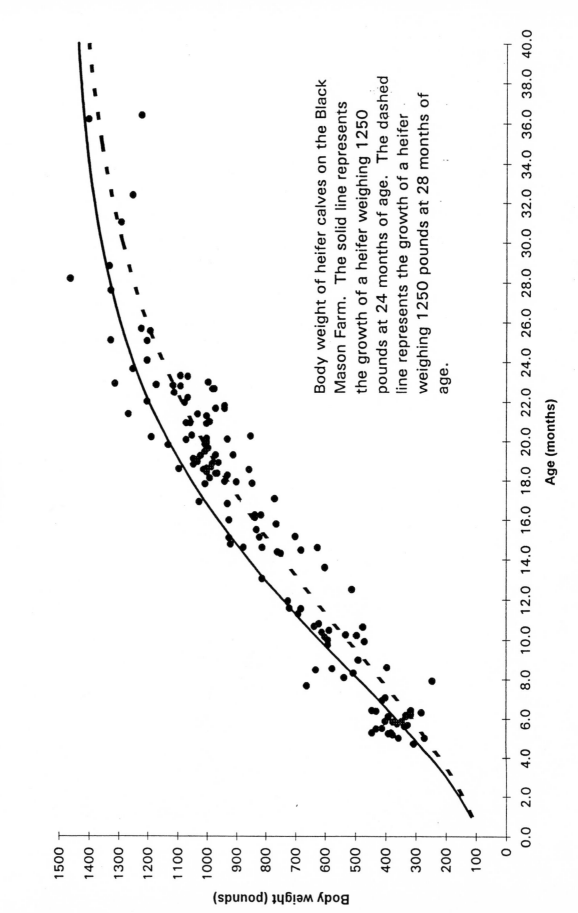

Body weight of heifer calves on the Black Mason Farm. The solid line represents the growth of a heifer weighing 1250 pounds at 24 months of age. The dashed line represents the growth of a heifer weighing 1250 pounds at 28 months of age.

EXHIBIT 4

Distribution of age at first breeding for heifers bred at least once and distribution of current age of heifers greater than 10 months of age and never bred on the Black-Mason Farm.

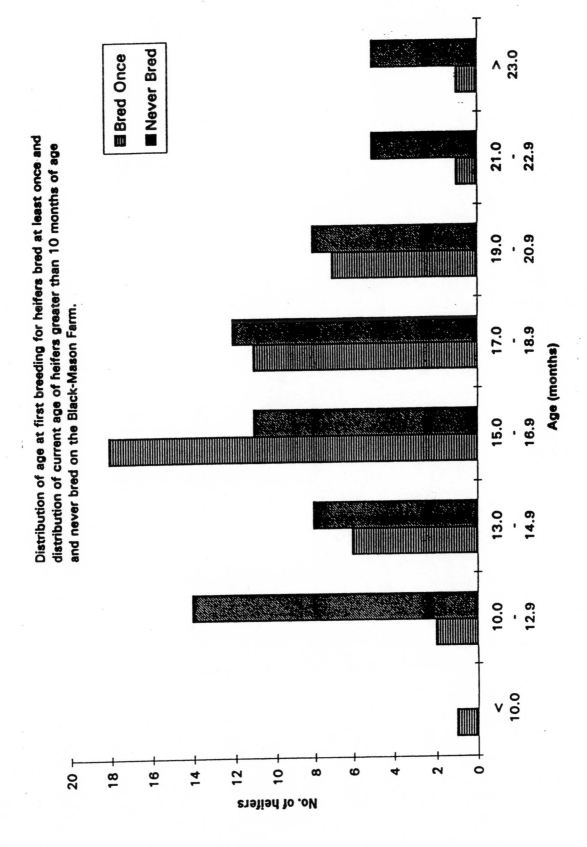

EXHIBIT 5

Partial Budget for Calving Heifers at 32 vs. 24 Months of Age

Prepared by Elmer Black

Decreased Expenses

Feed for Heifers

2,500 lbs. corn silage/day @ $25/ton	$ 32.25
2,500 lbs. haylage/day @ $55/ton	68.75
Mineral @ $0.04/calf/day x 93 heifers	3.72
Total	$ 104.72
$103.72 ÷ 93 heifers = $/heifer/day	1.13
TOTAL ($1.13 x 240 days = $/heifer)	**$ 270.25**

Increased Revenues

Milk

240 days x 70 lb./day = 16,800 lbs.

Total (16,800 lb. x $0.12/lbs. = $/heifer)	$2,016.00

Calves

$125/calf x 240 days/365 days x

0.5[a] = $/heifer of increased calf sales	41.25
TOTAL	**$2,057.25**

Increased Expenses

Feed for milking heifers

$3.85/cwt. x 168 cwt.	$ 646.80

Milking labor

3 min./day x 2 milkers x $0.11/min. x 240 days	158.40

Breeding

$37.50/pregnancy x 240 days/365 days	24.75

Labor

6 hours/200 cows x $6.50/hours x 240 days	46.80

Vet and medicine

$120/cow/year x $0.75	966.75
TOTAL	**$966.75**

Net profit/loss per heifer ($270.25 + $2,057.25 – $966.75)	$1,360.75
TOTAL PROFIT/LOSS ($1,360.75 x 93 heifers/year)	**$126,549.75**

---

[a] Assumes 50% of calves are bulls.

Thistledown Farms

*By Peter Tiffin and Oran B. Hesterman,
Michigan State University*

Dean Berden had farmed organically for nearly nine years and could not imagine ever using chemicals again. "Even though I've been driving a tractor for 43 years, if I had to return to chemical farming or quit farming, I would quit without hesitation." Dean knew that his way of farming was the way people should farm and that his knowledge of organic farming and organic food was very valuable to himself and others. Although he had already spent considerable time talking about organic farming, his biggest wish was to spend more time educating farm- and non-farm people about the virtues and advantages of organic farming for people and the earth. Not only was this Dean's biggest wish, it was also Dean's biggest dilemma.

History of Thistledown Farms

Nine years ago, in 1985, Dean and his father were farming 1,200 acres of corn, wheat, soybeans, and pasture using herbicides and synthetic fertilizers. Back then, it was Dean's job to apply the chemicals and fertilizer. His father had developed allergies and could not work around the chemicals. During the few weeks each spring that Dean applied chemicals he would feel "ornery and tired." Dean believed that the chemicals caused this reaction.

Dean also believed that farm chemicals were responsible for the deterioration of the health of the soil and the people around him. "I see people all around me getting cancer and other illnesses," he said. "An older lady who lives down the road has soft tissue melanoma. When she went to Chicago to have a doctor perform some tests, that doctor knew from her health problems that she must live in a farming community. He didn't even have to ask. The chemicals farmers use are doing this. I'm too concerned for my family to ever use chemicals again."

The negative effects Dean and his father attributed to chemicals motivated them to experiment with reduced chemical use. Their first experiments were not anything radical, mostly just reducing recommended rates of herbicides. In 1983, they decided not to use herbicides on one of their corn fields. Dean admits to being nervous and losing some sleep that year. However, weed pressure in that field ended up no worse than in fields where they had applied herbicides. That success helped Dean overcome his biggest obstacle to farming without chemicals – his own attitude. Once Dean decided he could farm without chemicals, the cultural and mechanical innovations and adaptations necessary were not difficult.

The Farming Operation

In 1994, Dean, in partnership with his father but with no hired help, farmed 560 acres of organically certified land near Snowden in the "Thumb" of Michigan. In 1988, Dean's farm was one of the first three farms in Michigan to be certified organic by the Organic Crop Improvement Association (OCIA). To maintain organic certification, Dean was not allowed to apply any synthetic fertilizers or chemicals to his land or to the grains he produced (Exhibit 1). An OCIA representative visited Dean's farm annually to look for evidence of chemical use and interviewed Dean about his farming practices.

Without synthetic fertilizers and chemicals, Dean depended upon crop rotation and mechanical means to maintain soil fertility and control weeds. His crop rotation consisted of soft white winter wheat — alfalfa or red clover — and one or two years of dry beans. In any one year, Dean had approximately 180 acres in alfalfa or red clover, 180 acres in winter wheat, and 200 acres in dry beans. Dean described the legume green manure crops as the lifeblood of his farming operation; fixing nitrogen from the air, providing potassium and phosphorous to the system, suppressing weed growth, and keeping the soil healthy. The wheat was used as a companion crop for establishing the legumes and helped minimize weed populations by shading out some weeds. Dry beans were also a good crop for Dean since these legumes fixed most of the nitrogen they needed.

Timely Cultivation Crucial

Although Dean's crop rotation helped to minimize weed pressure, mechanical weed control was necessary. Dean used three 12-foot cultivators, two 30-foot rotary hoes, and one 20-foot Lely Weeder. The weeder had a series of tines which slightly moved the surface soil, killing weeds when they were very small. Dean said that weeders were common in the days of his grandfather but were seldom used today. For improved cultivation, funnel shields and special teeth had been added to the rotary hoe, and shields had been added to the weeder. In addition, a row-marking system on all of his equipment allowed Dean to cultivate accurately within "three fingers around the row".

Dean placed such importance on the timing of cultivation that he personally inspected each of his fields daily during the three to four weeks before and after bean planting. The importance of proper cultivation was also why Dean refused to farm more acres than he could cultivate in a single day. As Dean said, "timing is 99 percent of organic farming." Proper timing of cultivation involved using the proper cultivation tool at the proper growth stages of both the crop and weeds. Learning proper timing required years of observation, and trial and error. After nine years, Dean's fields often looked less weedy than his neighbors', even though his neighbors made several applications of herbicides.

The Processing Facility

The OCIA requirement that organically certified products meet organic standards during processing as well as production prompted Dean to build a cleaning and packaging facility for beans and grains. In 1994, Dean had the only processing facility in the state certified by OCIA as a corporate facility. The corporate certification allowed him to process products produced on other farms without loss of organic status. The demand for his facility and services was high and he was cleaning and packaging grains year-round. In 1993, Dean and one full-time employee processed more than 2,250 tons of wheat, dry beans, and soybeans. The bulk of

these products were from Michigan, but Dean also received shipments from Canada, Ohio, and North Dakota.

Dean was deeply committed to the philosophy and practices he used in producing food. He thought he was farming better than his neighbors because he didn't apply toxic chemicals to his land or the food he produced, had good yields (Exhibit 2), and made money. Dean liked to describe himself as a producer of "healthy food." He believed that his practices were healthy for him, for the environment, and for those who ate the food. Organically produced food, according to Dean, was more healthful because it had a greater mineral concentration than conventionally produced products and didn't poison the consumer with toxic chemicals. Dean was also proud of the prices he received for his products. "The neighbors took notice when they saw the check I received for selling $17 soybeans."

Self Taught and Sought Out

Much of what Dean and his father knew about farming without chemicals they had learned by themselves. When they first started farming organically, there was little information available on chemical-free farming. Dean had contacted his county Extension agent, but the agent actually laughed at him, saying he was crazy to even want to try to farm without chemicals. Dean's contacts with specialists at Michigan State University weren't any more helpful. He was convinced that MSU and Extension had neither the expertise nor the desire to educate farmers and consumers about organic production and organic food. Without these more formal channels of education, it was up to farmers and farm organizations to provide information about organic food. Dean had taken part in this educational process as a member of both the Thumb Chapter of the Organic Growers of Michigan (OGM) and the Michigan Chapter of OCIA, and as a full-time high school woodshop teacher.

Several people had asked Dean for advice and information on his farming practices. Some were young farmers who were interested in reducing inputs. Others were farmers who had gotten into financial trouble and were hoping that reduced-input farming would be more economically viable. Dean had even been asked about his methods by the president of the local bank. Apparently, the bank president was interested in whether some of Dean's practices were worth recommending to other farmers. Dean saw more farmers becoming interested in organics as they heard about potential economic advantages. The few farmers who had followed Dean's path to chemical-free or reduced-chemical practices were a source of pride to Dean.

Sharing the Message of "Organics"

In the past, Dean had hosted farm field days in conjunction with the Thumb Chapter of the Organic Growers of Michigan. These field days attracted farmers and consumers to his farm where they saw his practices and learned the benefits of organic foods. These field days were small festivals—organic beans were cooked and served, and people sold organic products. The field days were very popular, attracting approximately 350 people in 1991 and nearly 450 people in 1992. Although Dean would rather not have so many people wandering around his farm, he thought that the damage and trouble were worth the education that resulted. In 1993, Dean helped organize the farm day but didn't host it because "it was time for some of the other farmers to show off their farms."

In addition to field days, the OGM-Thumb Chapter had a booth at several local fairs including those in Sanilac, Tuscola, and Imlay City. These local fairs gave Dean and most of the other members of the Thumb Chapter an opportunity to spend a day at the booth talking to a lot of people. Dean said there was a lot of interest expressed at these fairs. People would come up and say, "Hey, I've heard about this organic stuff, just what is it?"

More recently, several Thumb Chapter members had operated stands at the Royal Oak and Pontiac farmers' markets. The response at these markets had been terrific, offering the growers an opportunity to both market their products and educate the public.

Dean thought that other opportunities existed for educating more people about organic food. Greater education would result in a larger, more diverse and healthier organic foods community, he reasoned. Some of these educational possibilities might include consumer education weekends for families who would pay $10 to $15 to be bused to various farms and spend the night camping together. This would give them a chance to share organically grown food and discuss the importance of organic farming and what it means for individuals, the environment, and communities. Another idea was to find some time during the summer for day-long seminars in which Dean would invite people to his farm to see and learn about his operation. Dean knew there were many people interested in organics and he wanted to help further their knowledge. He was particularly excited about the possibility of involving inner-city people and people of color in a predominantly white movement.

The Dilemma

Dean's greatest constraint was time. He already had a very full schedule which made him very selective about those with whom he spent his time. This was reflected during the summer of 1993 when Dr. Oran Hesterman, a professor at Michigan State University, called to ask if a group of students could come to tour Dean's operation. Rather than grant the request immediately, Dean asked Dr. Hesterman first to send a list of those who would attend and an explanation of why they wanted to come. Dean did allow the group to come for a tour, and in contrast to the initial conversation, he spent several hours showing the group around and entertaining questions.

Dean's commitment to his school, farming, and his family had forced him to turn down opportunities to talk and give tours. Most recently he had to decline an offer to speak to a church group from Flint. He was invited to speak to this group by a man who bought food from Dean and became a proponent of organic food after it apparently cured his Chronic Fatigue Syndrome.

Dean was so committed to the virtues of organic agriculture that he found it difficult to decline such opportunities. However, he realized he couldn't commit to everything. "After all, there's only 24 hours in a day and I've got to sleep, and my health and my family come first," he said.

Questions for Discussion

1. Can Dean expand his educational effort while continuing to successfully farm, run his processing facilities, teach, and spend time with his family?

2. Are there ways for Dean to find time to put more effort into education without sacrificing other activities?

3. Is there anything unique or admirable about Dean's farming system that people would or should be interested in learning?

4. Does Dean's knowledge about organic farming and organic food have value? If so, how can Dean realize the value in a tangible way?

5. Why does Dean want to spend more time educating?

6. What are Dean's ideas about how to educate people? Are these good ideas? Are there other ways that Dean could educate and share information?

7. How important is it that the consumer learn about how food is produced? Is it more important for someone like Dean than for a farmer using more traditional methods? Why?

EXHIBIT 1

Excerpts from the OCIA Information Packet

INTRODUCTION

Welcome to OCIA! The Organic Crop Improvement Association is the worlds largest certification program. OCIA membership can be found in North, South and Meso-America, Europe and Asia. It is a farmer owned and managed grass roots democratically run program that focusses on crop and process improvement for farmers, processors and manufacturers. Certification is seen as a tool for the improvement of the quality of foods grown in an environmentally and sustainable production system. Found in this information package is the basic information on how to become a member and what the basic requirements are for certification. Should you have any questions, please contact our International office.

MISSION STATEMENT

"We, producers, handlers and consumers of organic food and fiber build environmental stewardship through ethical partnership with nature." OCIA mission statement adopted at Michigan State University, December 1991.

OCIA'S BRIEF HISTORY

In the depression years of the dust bowl, farmers started getting together informally to share their mutual farming experiences. Having no technical support to enhance the development of their profession, they formed the first crop improvement associations. The principles were simple: farmers are the experts on their lands; having regular meetings as opportunities to share their experiences with some techniques and trials, they could acquire the basis of adapted technology.

In the mid seventies, the notion of organic agriculture began circulating within a group of pioneers. One of these people, Tom Harding, seeing a certain parallel with the technological situation of the nineteen twenties, started working on the concept of an organic crop improvement association: farmers getting together as a grass roots movement to facilitate the development and the transfer of technical expertise for organic farmers. In the early eighties, while he was working to develop organic quality control measures for a wholesaler on the U.S. East Coast, Tom started structuring the certification content of what was to become the Organic Crop Improvement Association. After two years of certifying for the wholesaler, chapters started forming independently and assuming the leadership of the program.

It was in the fall of 1985, in Albany (New York), that many individuals from different groups got together and structured the concept of a farmer owned and farmer controlled association. This first meeting expanded the vision and set the corner stone of what was to become one of the worlds largest organic certifiers.

In 1988, the OCIA program took on an international identity when a group of Peruvian farmers joined the program because of the concept of crop improvement and farmer networking. From then on the program began a tremendous expansion throughout Latin America. In the early nineties, membership from Asia and Europe added a new dimension to the international body.

Four chapters started certifying in 1986, representing less than sixty (60) farms: seven years later, in 1993, we are 65 chapters representing more than ten thousand farmers and two hundred corporate members from twenty five or so countries.

WHAT IS CERTIFICATION

When consumers of organic products go out to the farms to buy their foods, they establish a relationship based on their trust of how the farmer is producing. As organic foods move into a more structured marketplace, consumers must be reassured that the foods they are purchasing through retailers, wholesalers and processors are genuine and that the integrity of the initial farm product is maintained throughout the chain of custody. Certification is a set of procedures by which the organization verifies that the production system respects the strict standards of organic production. This process guarantees an impartial third party assessment of organic quality for foods bearing the OCIA trademark.

CERTIFICATION

INSPECTORS FEES AND TRAVEL COSTS

For those who are accepted as members and decide to pursue certification, our International office will work with the applicant to select an approved third-party certification inspector. We accept only OCIA approved independent agents and these agents contract directly with applicants. We will send copies of your preliminary certification questionnaire and any related material to the inspector.

Once the inspector has reviewed your questionnaire, he will call you to discuss the scope of the certification project. You should get a written estimate of the agents fees for travel time, on site work and travel expenses. You are responsible for all fees and costs associated with certification.

If you come to terms with the inspector, send a check in the amount of his/her fee to the OCIA International office. This check should be made out to the agent. Next send the agent a check to cover his estimated travel costs. On completion of the site visit, you may need to work with the agent on adjustments in fees and costs. OCIA will hold the inspectors fee until the site visit is complete and all the required reports are submitted to the International office.

OCIA CERTIFICATION REVIEW

The agent prepares a written report for the OCIA International Certification Committee. This committee is composed of organic farmers, processors, traders and retailers. All have many years of experience in one or more phases of the organic food industry. The committee will examine the certification agents report in light of current standards and then recommend or deny certification. You will receive a written summary of the review explaining the committee's decision.

CONFIDENTIALITY

OCIA wants to work closely within the organic food industry to help coordinate crop improvement, certification and audit trail systems. We fully recognize the sensitive and often proprietary nature of the information in this industry. As a result, the names of those seeking certification are held in strict confidence by the third-party agent, our administrative staff and the International Certification Committee.

Since Certification Committee members are from the organic food industry, some applicants may be concerned that sensitive information will be compromised. These applicants are encouraged to ask the OCIA Administrative Director for the name of alternate Committee members. The alternate will stand in for the member-in-conflict during the certification review process.

HOW CAN OCIA CERTIFICATION BE CONFIRMED

Each OCIA certified grower and processor is issued a certificate with a unique number. Chapters issue certificates to their members while the International office issues them to At-Large and Corporate Members.

Simple transactions involving OCIA certified products should be accompanied by the growers OCIA number. Where products of many OCIA producers are mixed, the audit control system must trace these products to the processors and growers involved.

Proof of certification can be confirmed through transaction certificates (See CMC Section). Our members are proud of our certified organic products and our production, processing and audit trail systems. Certificate numbers can be checked against chapter and International office records to confirm authenticity.

THE OCIA TRADEMARK SYSTEM

The OCIA trademark is a farmer-owned seal of quality for organic foods, backed by an audit trail, which can trace a product from the market shelf, back through all the intermediaries (including the farmer), right to the seed. Such an audit trail is the consumers best guarantee that a product really is organically grown and processed, but it is also valuable at all stages of the production, processing, and distribution cycle as a form of protection against liability exposure for problems caused by someone else.

The OCIA CERTIFIED ORGANIC trademark is brand-neutral, which means that trademark license holders have the right to sell to whatever market brand they wish. Consequently, one can find several different brands of the same product, all certified by OCIA, and all competing with each other. This competition is based on market factors, rather than on the idea that one product is somehow "more organic" than the other. In this way, the OCIA trademark is similar to the REAL dairy products trademark, the Woolmark (pure wool), the Underwriters Laboratory seal, the Kosher K, or other marks which attest that the products to which they are affixed are genuine, safe, or otherwise produced according to the uniform and consistent standards.

The Organic Crop Improvement Association is a confederation of groups and individuals all over the world. Each Chapter, Member-At-Large, or Corporate Member is represented at the confederation, which governs all matters pertinent to the group as a whole, such as minimum standards, trademarking, and membership policies.

The Audit Control System is one of the most important parts of the OCIA Program and is composed of the Certification System. Internal and External Review Process, Certification Agent Review and Registry, and the complete On-Farm and Processing and Distribution Production, Sales and Inventory Control Recordkeeping System that extends to the Retail Market System and Product Labeling.

It is through this rigid control system that we endeavor to authenticate to the consumer the OCIA Product Quality Guarantee:

OCIA PRODUCTS ARE FOOD YOU CAN TRUST!

LICENSING

Licenses are renewable annually, dependant upon recertification, and all trademark users must pay a user fee based on their annual sales. Sometimes, a processor or wholesaler pays the certification fees and user fees for a farmer or other supplier. In this case, the right to use the trademark belongs to the processor or distributor. The farmer may still sell to anyone else, but these products may not bear the OCIA trademark. Processors or Distributors may request any supplier to sign a normal sales contract with an exclusivity clause, but this is not related to the trademark. No person or company can demand or claim an exclusive right to all OCIA certified products in a particular commodity. Consequently, OCIA certified and trademarked products are in free and open market competition, backed up by consistent high standards, independent third party certification, and an extensive program of organic crop improvement. This is the consumers best guarantee of a top-quality organic product at the best price.

USING OCIA CERTIFICATES AND AUDIT TRAILS TO SELL OCIA CERTIFIED PRODUCTS

The active use of OCIA transaction certificates and audit trails in the marketplace will help distinguish OCIA certified products from the ordinary. As OCIA members, we offer our customers much more than personal health benefits.

Although personal health is very important, we should also be telling our customers about our efficient biological farm management and business management practices. Consumers need to know they can work with us to eliminate costly farm subsidies, harmful agricultural chemical and soil erosion. Consumers should also know that our farms and ranches offer abundant food and cover for all kinds of wildlife.

By supporting OCIA certified brands, consumers support a unique, producer-controlled organization with corporate members from all phases of the organic food industry. As OCIA certified farmers and ranchers, we should tell consumers that OCIA corporate members actively support sustainable agriculture. This support is in the form of regular purchases of OCIA certified farm products at premium prices.

Organic Crop Improvement Association

1994 International Certification Standards

As Revised December 1993

Organic Crop Improvement Association International

3185 Township Road 179
Bellefontaine, Ohio 43311
U.S.A.

Telephone: (513) 592-4983 • Fax: (513) 593-3831

FARM CERTIFICATION STANDARDS

2.1. ADMISSIBILITY

2.1.1. Certification may be on a whole farm or on a field by field basis. If the latter, all fields of the farm unit must be committed to an ongoing program of organic crop improvement. This program of organic crop improvement must be submitted in writing to the certification committee by the third year of certification and be designed to bring 100% of the farm into transition within at least 5 years following the first certification of any portion of that farm. Exceptions may be granted for portions of the farm which are rented or not under the complete control of the grower or for unexpected and extreme circumstances. Fields may not be rotated in and out of organic production and remain certified.

2.1.2. No crop can be sold as "OCIA Certified Organic" if the same crop is also produced elsewhere on the farm using methods or materials that do not conform to these standards, unless the farmer can clearly demonstrate that there exist both the physical facilities and the organizational ability to ensure that there is no possibility of crop mixing. This criterion applies equally to situations when uncertified crop is produced by the same farmer on another farm unit, or is purchased for resale.

2.1.3. A field can be certified organic if there has been no use of non-acceptable materials (insecticide, herbicide, fungicide, fertilizers, etc.) or methods during the three years before harvest. This time frame is set to confirm the applicant's commitment to organic practices.

2.1.4. In cases where an adjoining farm is growing heavily sprayed crops, or there is other possibility of contamination, there must exist adequate physical barriers or a 25 foot (8 meter) minimum distance between organic crops and sprayed crops to maintain the integrity of certified fields. When contamination is suspected, the certification agent shall require residue testing. Residue levels must not exceed the lowest legal organic standards existing in any country where the OCIA seal is used.

2.1.5. Complete information describing at least three (preferably five) most recent years' production methods and materials, as well as information about current production practices, must be provided. The applicant for certification must also furnish an outline of farm management strategies directed at achieving strict compliance with these standards.

2.1.6. To be certified, a farm or field must be managed in accordance with the following required practices listed below, using authorized methods and materials.

2.2. AGRICULTURAL PRODUCTION

2.2.1. REQUIRED PRACTICES

2.2.1.1. Development and implementation of a conscientious soil building program designed to enhance organic matter and encourage optimum soil health.

2.2.1.2. Rotation of non-perennial crops in accordance with accepted regional organic practices.

2.2.1.3. Use of careful management, resistant varieties, intercropping, and maintenance of soil health as the first line of defense against weeds, pests, and diseases.

2.2.1.4.   Generation of an audit trail which will permit tracing the sources and amounts of all off-farm inputs, date and place of harvest, and all steps between harvest and sale to the wholesaler, retailer, or final consumer. Certification agents shall recommend denial of certification for inadequate audit trailing.

2.2.1.5.   Maintenance of machinery and equipment in good enough condition to avoid contamination of soil or crops with hydraulic fluid, fuel, oil, etc.

2.2.1.6.   Use of pre- and post-harvest handling procedures and packaging materials which ensure maximum product quality (appearance, hygiene, freshness, and nutrition) using techniques and materials that are consistent with these standards. Irradiation of certified foods is prohibited.

2.2.1.7.   Beginning March 1991, soil testing will not be mandatory for inspection and/or certification purposes. However, OCIA maintains that all members manage their soils responsibly, with the intent to improve soil fertility and tilth through proper management practices. If any problems arise that are associated with nutritionally imbalanced soils, such as poor plant growth or excessive pest pressure (including insects and/or weeds), then it is the responsibility of the grower to test the field(s) in question for macro- and micronutrients, cation exchange capacity, base saturation, and organic matter. These results should be used in part to determine reasonable management options to correct soil imbalances and improve field and crop performance. Further, it is the inspector's responsibility to document field problems associated with nutritionally imbalanced soils, and to verify whether soil tests have been taken and/or soil test results received and appropriate action has been taken. In this case, appropriate action may include developing a more appropriate rotation, applying an OCIA approved material, modifying composting and/or manure management practices, or responding in some other manner to the acknowledged problem. Failure to respond in any manner to a known soil deficiency(s) that result in inferior quality crops and/or poor soil quality will be looked upon as negligence in management and may be used as grounds for de-certification.

2.2.2. SOILS AND PLANTS

Authorized Methods and Materials

2.2.2.1. Organic Matter

   a.   Composted manure, preferably produced on the farm, or if imported which is free of contaminants.

   b.   Uncomposted manure that has been turned and free of internal frost for at least six months prior to application.

   c.   Fresh, aerated, anaerobic, or "sheet composted" manures on perennials or crops not for human consumption, or when a crop for human consumption is not to be harvested for at least four months following application. At application the soil must be sufficiently warm (about 10 C) and moist to ensure active microbial digestion.

   d.   On radishes, leafy green, the beet family, and other known nitrate accumulators fresh, aerated, anaerobic, or "sheet composted" manures may not be applied less than four months before planting. At application the soil must be sufficiently warm and moist to ensure active microbial digestion.

   e.   All manure sources and management techniques must be clearly documented as a part of the certification process.

   f.   Green manures and crop residues, peatmoss, straw, seaweed, and other similar materials. sewage sludge and septic waste is prohibited.

g.   Composted food and forestry by-products which are free of contaminants.

2.2.2.2. Minerals

a.   Agricultural limestone, natural phosphates, and other slowly soluble rock powders. Fluorine content of the natural phosphates should be balanced with application rates so that total fluorine applied does not exceed an average of 5kg/ha/year in the field, or 10 kg/ha/year in the greenhouse.

b.   Wood ash, langbeinite (sulpomag), non-fortified marine by-products, bonemeal, fishmeal, and other similar natural products.

c.   Cottonseed meal, leathermeal, and blended products containing these substances are permissible only if free of contaminants.

d.   Highly soluble nitrate, phosphate, and chloride nutrient sources, natural or synthetic, are prohibited from use on soil or foliage.

e.   Ammonia and urea products are prohibited.

f.   Potassium sulfate (borax, Solubor), sodium molybdate, and sulphate trace mineral salts are permitted where agronomically justified. Application rates and distribution should be controlled by applying these products in solution with a well-calibrated sprayer.

2.2.2.3. Foliar

a.   Liquid or powdered seaweed extract or other non-fortified marine by-products. (Explanatory note: In some circumstances such as the use of phosphoric acid to hydrolyse fish emulsion a normal aspect of the industrial process coincidentally furnishes plant nutrients. This is not to be considered "fortification" for the purposes of these standards. The operative criterion is whether a product is added to the process in order to boost the analysis, as is the case with potassium nitrate added to fish emulsion.)

b.   Plant or animal based growth regulators and other plant or animal products.

c.   Adjuvants, wetting agents, and the like.

d.   Mineral suspensions such as silica.

2.2.2.4. Seed, Seedlings, Grafting and Root Stock

a.   Horticultural crops and non-perennial field crops must be produced from seed that has not been treated with any unauthorized product. Temporary exceptions can be made if untreated seed is not available.

b.   Annual transplants must be grown according to OCIA standards. Perennial transplants may be from any source, but crops sold as certified organic must be from plants which have been under organic cultivation for at least 12 months prior to harvest.

c.   Vegetatively propagated plants such as garlic and other bulbous plants are to be considered as seeds and fully respect paragraph [a.] of this article.

2.2.2.5. Other

a.   Assorted plant and /or animal preparations, biodynamic preparations, microbial activators, bacterial inoculates, and mycorhizae, etc.

b.   Microbes used in the production of certified crops or products must be naturally occurring (not the result of genetic engineering).

2.2.3. PEST CONTROL

Authorized Methods and Materials

2.2.3.1. Disease

a.   Use of resistant varieties.

b.   Lime-sulfur, Bordeaux, elemental sulfur. Other sulfur or copper products may be approved by the certification committee with the approval of OCIA.

c.   Fungicidal and cryptocidal soaps, plant preparations, vinegar and other natural substances.

2.2.3.2. Insects and Similar Pests

a.   Use of resistant varieties and the provision of conditions favoring natural equilibrium.

b.   Insecticidal soaps and botanical insecticide such as ryania, sabadilla, etc.

c.   Rotenone, pyrethrum, dormant oil (preferably vegetable-based), and diatomaceous earth may be used with great caution due to their high ecological profile.

d.   Sexual, visual, and physical traps.

e.   All pesticides containing aromatic petroleum fractions or synergists (such as piperonyl butoxide) are prohibited.

f.   Microbial insecticides as found in the OCIA material list are acceptable.

2.2.3.3. Weed Control

a.   Weeds are to be controlled through a combination of cultural practices which limit weed development (rotation, green manure, fallow, etc).

b.   Mechanical, electrical, and thermal wedding.

c.   Microbial weed killers.

d.   Chemical or petroleum herbicides are prohibited. Amino acid herbicides have not yet been registered for use.

e.   Use of plastic mulch will be subject to approval of the chapter.

Section Nine

9.0 Administration

9.1. CERTIFICATION PROCEDURES: INSPECTORS

9.1.1. The third party certification inspector is to be demonstrably impartial and independent evaluator of member compliance with these standards or those of the chapter to which the member belongs.

9.1.1.1. The inspector shall not be a party to any transaction involving the certified products.

9.1.1.2. The inspector may not be an employee of or have any financial interest in any company which is a party to any transaction involving the certified products.

9.1.1.3. Advice provided by the inspector shall be limited to helping the member meet standards and improve organic production techniques. Consultation for an additional fee at any time within the certification year is unacceptable and constitutes grounds not only for dismissing the inspector, but for revoking the member's right to use the seal.

9.1.1.4. The inspector shall not have worked for the applicant member in any capacity in the year prior to the certification year, and shall not work for the applicant member in the year following the certification year.

9.1.2. In cases of suspected contamination, or following a request from the certification committee, the inspector shall have the right to make unannounced visits, take samples, and require residue tests, all at applicant expense.

9.1.3. The relationship between certification inspector and member is one of confidence in all matters not pertaining directly to certification. In certain cases it may be necessary for the inspector to be bonded. It is also advisable for the inspector to carry liability and/or errors and omissions insurance.

9.1.4. Only OCIA approved inspectors will be authorized to do inspections for OCIA certifications.

9.2   CERTIFICATION PROCEDURES: CHAPTERS

9.2.1.   Chapters shall have a certification committee which consists of at least 50% farmers (a minimum of three should be OCIA certified unless the chapter is in its first year of certification), and includes two members with no financial interests in the production or marketing of product.

9.2.2.   The Committee shall:

9.2.2.1..   Define and implement standards; verify adherence to standards through peer evaluation and a third party certification inspector; ratify or reject the certification inspectors recommendation to certify or refuse to certify member farms.

9.2.2.2..   Administer the certification program including hiring the inspector, scheduling visits, coordinating paperwork, and ensuring that all requested documents are forwarded to Confederation offices.

9.2.3.   The certification inspector, shall, before the harvest begins:

9.2.3.1.   Visit at least one third (1/3) of the total fields on the farm and at least one field of each crop to be certified on the farm and verify that practices conform to these standards and to written information in the application. The fields to be visited are to be picked at random at the discretion of the inspector.

9.2.3.2.   Examine post-harvest handling facilities, evaluate the applicant's management skills and organizational ability, inventory materials, and ensure that equipment available for weed control, etc. is capable of doing the job required at the scale proposed.

9.2.3.3.   Discuss potential problems and possible solutions with an emphasis on product quality, audit trailing, and organic crop improvement.

9.2.3.4.   Meet with the certification committee and upon request recommend to certify, or to refuse to certify an applicant member.

9.2.3.5.   The first year, that certification is granted only crops harvested after the inspection visit are eligible for certification status. For farms being re-certified the members certification will be considered valid for a full year following the chapter certification committee's first decision in favor of certification. A 30 day grace period may be added to this year by the committee for problematic situations if certification review is in process. In the case of re-certification no more than 75% of crops may have been harvested at the time of inspection.

9.2.4.   The chapter shall sign a seal licensing agreement with the pertinent national OCIA corporation, and shall grant rights to use the OCIA Certified Organic seal to certified members in accordance with normal seal control procedures. The chapter shall ensure that seal use complies with normal or accepted OCIA practices.

9.2.5.   Only OCIA approved inspectors will be authorized to do inspections for OCIA certifications.

9.3.   CERTIFICATION PROCEDURES: MEMBERS AT LARGE

9.3.1.   Certification procedures for members at large are identical to those for chapters ,with the exception that the confederation's Certification Review Committee replaces the certification committee of the chapter.

9.4.   APPEALS

9.4.1.   An appeal may be initiated against either a refusal of certification or the granting of certification.

9.4.1.1.   Any member or applicant may initiate an appeal, even against a decision made in another chapter.

9.4.1.2.   Burden of proof is on the party initiating the appeal.

9.4.1.3.   Expenses will usually be borne by the losing party to the appeal.

9.4.2.   Appeals of a certification committee decision shall be heard by an ad hoc tribunal consisting of one member of each of three neighboring chapters, duly appointed by the IRC, provided none is a party to the appeal.

9.4.2.1.   In appeals against certification, the grower should be notified of the complaint and its nature, be furnished with an outline for response (audit trail, farm plan, financial books, etc.), and respond within 72 hours indicating whether the appeal will be contested.

9.4.2.2.   The appeal tribunal shall hear arguments within ten days, and may seek amicus curiae submissions from others. The tribunal decision shall be final and binding for the certification year.

9.5.   AMENDMENT PROCEDURES: CERTIFICATION STANDARDS

9.5.1.   Standard proposals and amendments pertaining to agricultural crop production are subject to review at the OCIA Administrative Council meeting which is held in the last quarter of each year. Standard proposals and amendments pertaining to animal husbandry and food processing are subject to review at both the OCIA Administrative Council meeting which is held in the last quarter of each year and the Administrative Council meeting which is held in conjunction with the Annual meeting the first quarter of each year. Proposed change in standards must be submitted by any member including chapters, farmers at large or corporate members to the Standards Committee at least 60 days prior to the Administrative Council meeting. The Standards Committee can also formulate amendments for adoption by the Administrative Council. The Standards Committee shall mail a copy of proposed deletions or additions to the last recorded address of each OCIA Associate at least 45 days before the Administrative Council meets to consider their changes as referred to in the by-laws.

9.6.   EXTERNAL CONTROL

9.6.1.   The Certification Review Committee of the Confederation shall appoint a demonstrably independent third party agency to verify at random that the certification control procedures of chapters meet a consistently high and uniform professional standard. Section Ten

EXHIBIT 2

Crops, approximate yields, and prices from 1993.

CROP	YIELD	PRICE
Soft white winter wheat	75 bu./acre	$6.50-8.50/bu.
Navy beans	2000 lb./acre	$0.45-0.65/lb.
Black turtle beans	2000 lb./acre	$0.45-0.65/lb.
Adzuki beans	1200 lb./acre	$0.85-1.20/lb.
Soybeans (Vinton)	32 bu./acre	$18.00/bu.

The Boundaries of Chemical Drift

By Amy Roda and Oran B. Hesterman,
Michigan State University

As Elizabeth Bergson watched the Agrochem spray truck round the last bend of her neighbor's field, she shook her head in frustration and in anger. "Go back home and shut your windows," she muttered sarcastically. "Just because they are behind schedule, they have put my health and my orchard at risk."

Early on the morning of May 16, 1993, Elizabeth had awakened to a noxious smell drifting in through the window. Upon investigation, she found a commercial chemical company applying herbicides to the neighbor's no-till corn field. Looking up at her weather vane, she realized what had created the smell. "The wind was blowing at about 25 mph, carrying herbicides across the road to my farm."

Managing the Organic Orchard

Elizabeth owns and manages a 17-acre fruit orchard in Michigan. In 1987, she took over the orchard from her uncle who had grown apples on the land for 24 years. That same year, she began the process of converting the orchard to organic production. In 1990, the farm received organic certification from the Lifeline Chapter of the Organic Growers of Michigan (OGM) (6) and Organic Crop Improvement Association (OCIA) (9).

Elizabeth's yields are about average for her area of the state (400 bu./acre). Usually, about 15 percent of the yield is sold as whole fruit for the table. Elizabeth processes approximately 70 percent of the apples into cider and the remaining 15 percent is made into apple butter which she sells under her own label. These, and other organic products produced by area farmers, are sold at the farm's store. Her largest buyers are food cooperatives located in Lansing and Ann Arbor. Through these outlets, Elizabeth makes a healthy profit (Exhibit 1). Compared to local orchards, she receives about

34 percent more than conventional prices for table apples, 50 percent more than conventional prices for cider, and similar prices for apple butter.[1] These differences in revenues help offset the 23 percent greater costs she incurs producing fresh organic apples compared to similar sized conventional orchards (4). Her processing costs for cider and apple butter are comparable to other non-organic orchards.

Pesticides and the Public

Elizabeth enjoys visitors. Each year she participates in Mid-Michigan Harvest Trails, a self-guided tour of the region's farms. Visitors are welcome to look around the orchard and cider mill and learn organic "secrets" from Elizabeth. From her interaction with the public, she senses a growing public anxiety about pesticide residues and additives in food. She stresses that organic farming is not just free of synthetic chemicals, but signifies a way of working with the natural systems that build the soil.

Elizabeth is aware of the public's limited view of organic farming. She remembers times when visitors (who had just proclaimed the benefits of a chemical-free system) winced when they saw that the apples for sale at her farm were not "wax perfect," and chose to purchase the apple butter or cider instead. Elizabeth says her secret to successful organic farming is "keeping your plants and soil healthy as the key to controlling pests and to good production." She fertilizes the orchard with rock phosphorous, green sand, trace minerals, fish and seaweed emulsions, vinegar, and cow manure. She manages weeds by mowing. Through regular scouting, Elizabeth determines when insect control is necessary. She applies botanical insecticides, including rotenone and pyrethrum, as well as BT (*Bacillus thuringiensis*), vegetable grade hydrogen peroxide, and vinegar as needed (7).

---

[1]Percentages based on conventional prices of $7.50/bu. of table-quality apples, $2.00/gal. of cider, and $0.194/oz. of apple butter.

Drifting Herbicides

Elizabeth reacted immediately on that spring morning once she realized that the herbicides sprayed on the no-till field were drifting over to her orchard. She first approached the Agrochem, Inc. operator, and informed him that she could smell the herbicides across the road and asked him to stop spraying. He advised her to "go home and shut the windows." Riled, Elizabeth insisted that he stop at least until the wind died down. "Look lady, we are behind schedule. If you want me to stop, you'll have to talk to my supervisor," he responded. She returned home and called Agrochem, Inc. The supervisor's rebuttal echoed the operator's: "go inside and shut your windows, we should be finished shortly." Realizing that the supervisor was not about to order the operator to stop spraying, she contacted the Lees, owners of the no-till field.

While Elizabeth had chosen to practice organic methods of farming, the Lees were taking a no-till approach to reduce production costs and soil erosion. They were facing declining profits and were trying to keep their land economically viable. Elizabeth explained the chemical drift situation to Mrs. Lee. After glancing out the window, she told Elizabeth, "It must not be as windy as you think; the clothes out on the line aren't moving." Because they rented-out the parcel adjacent to the orchard, Mrs. Lee felt that they really had no say in the matter and suggested Elizabeth contact their renter.

Elizabeth called the renter, Mr. Rice. She found him in agreement that Agrochem probably should not be spraying. However, he did not offer to halt the spraying. Exasperated, she made a final call to the county agricultural Extension agent, John Gault. He contacted the company's supervisor with hopes of resolving the situation. The supervisor was surprised to learn that the orchard was a certified organic farm. He promised to radio the operator, but by this time, the spraying was almost completed.

From her farm, Elizabeth watched the operator finish the field and move on to his next job. Musing to herself, she considered what she could do to stop this casual exchange of chemicals between farms. "For my health or even the health of a jogger, this should not be happening," exclaimed Elizabeth. Elizabeth never felt herself part of the "inner circle"

of area farmers. Not knowing if she had legal recourse, she wondered if any further "meddling" would rile her neighbors. "But if I do nothing, will this happen again next year?" she asked herself.

Rules Governing Organic Certification

Until 1991, Elizabeth maintained both OGM and OCIA certification. For the 1992-93 season, Elizabeth chose to maintain her OGM organic certification and temporarily discontinue certification with OCIA. Organic certification from OGM was sufficient for products marketed in Michigan (6). Farmers interested in marketing their products outside the state, or internationally, usually acquired certification from OCIA (9). Elizabeth found that her Michigan market was expanding enough that she did not need to seek markets outside the state.

Both OGM and OCIA certification procedures require a minimum of three years waiting period after application of any prohibited fertilizer or pesticide, including herbicides (Exhibits 2 and 3). Farmers are requested to submit complete and accurate records including all crops, soil amendments, and soil tests. A visual inspection of the farm is made after review of the material. OGM usually sends a local representative who is familiar with the community, while OCIA requires a third party, often from out-of-state. A written report of the visit is reviewed by a committee that either grants or denies certification.

Each year, the farmer submits a renewal application and is inspected before recertification is granted. If the inspector discovers any noticeable signs of chemical adulteration, the farmer can be required to test the suspect areas for chemical residues. Any areas with known contamination are removed from organic status for three years. Products from these areas cannot be marketed as organic, but can be sold to conventional buyers.

Farmers are trusted to provide complete and accurate information to the inspector and certification board. Elizabeth wondered whether she was obliged to reveal the chemical adulteration, or whether her field would test positive for contamination in the next inspection. Either way, she was in jeopardy of losing her organic certification.

Chemical Weed Control in No-Till Corn

On May 16, 1993, Elizabeth's orchard was in an early stage of leaf development (1, 2). Blossom buds were visible but they had not begun to open. Inspecting the orchard that evening, Elizabeth found no visible evidence of chemical damage to the apple buds or the leaves or to the grass surrounding the trees. She also noticed that the weeds growing along the roadside did not show any signs of wilt or discoloration.

John Gault, the Extension agent, had access to a guide listing the herbicides generally used to control weeds in a no-till corn system (10). Michigan State University Extension (MSUE) crop specialists are knowledgeable of visual symptoms of pesticide drift onto off-target plants (3). Elizabeth is aware that Extension provides weed control information to farmers but has not used these resources.

Regulation of Pesticide Use

All commercial applicators operating in Michigan must be certified by the Michigan Department of Agriculture (MDA). Applicators must successfully complete two written examinations based on information provided in a training manual developed by MSUE (5). The manual addresses the general standards required of all commercial applicators. Each county office facilitates certification seminars for both commercial and private applicators. Michigan laws that govern commercial pesticide applicators and their pest management operations are described briefly in the manual (5). MDA distributes copies of the written laws to each certified commercial applicator.

Elizabeth's orchard is registered with MDA as a certified organic farm. MDA distributes the list of organic farmers to all certified commercial chemical applicators. The MSU county Extension office also keeps a record of organic farmers.

Elizabeth's Decision

After the events of the morning of May 16, Elizabeth was faced with a difficult decision. The organic farm that she had spent years developing was placed in jeopardy by chemical drift. This danger was far more than economic for Elizabeth; she believed that chemicals represented a real danger to her personally and to everyone else in the area, from landowners to the passing jogger. Because she was dedicated to education and progress in organic farming, she felt some obligation to file a lawsuit to prevent the situation from happening again to her, or to someone else.

Elizabeth also knew that chemical drift was accepted by farmers as a natural part of the business. She was convinced that she would not find any support from her neighbors during a litigation process. She did not know who was responsible for the spraying, or what her legal rights were. Would taking legal action be worth the costs, potential loss of organic certification, and possible alienation from the community? And to whom should this action be directed?

Study Questions

1. How should the extent of chemical drift be determined and by whom?
2. What effects (long-term and short-term) could herbicides commonly used in no-till corn production have on the orchard?
3. If found, how would chemical residues affect organic certification? What economic impact would this have on the orchard?
4. What precautions could Elizabeth Bergson and Agrochem, Inc. have taken to prevent the situation?
5. Should Elizabeth take legal action?
6. What should Elizabeth do?

References

1. Chapman, P.J. 1966. Standard names for key apple bud stages. Proc. *New York State Hort. Soc.* 111: 146-149.
2. Edson, C.E. 1986. "An apple phenology study: Design of a predictive model of shoot growth, flower development and fruit growth." M.S. thesis. Michigan State University, E. Lansing, MI.
3. Gunsolus, J.L. and W.S. Curran. 1991. "Herbicide mode of action and injury symptoms." North Central Regional Extension Publication 377. University of Minnesota, St. Paul, MN.

4. Kelsey, M.P. and P. Schwallier. 1989. "Cost of producing fresh apples in western Michigan." Extension Bulletin E-1107. Michigan State University, E. Lansing, MI.

5. Landis, J.N., R.R. Rosenbaum, and J.A. Stackecki. 1992. "Commercial and private pesticide applicator core manual: Certification and registered technician training." Extension Bulletin E-2195. Michigan State University, E. Lansing, MI.

6. Lifeline Chapter of the Organic Growers of Michigan Council. 1990. "Certification procedures and standards." Lifeline Chapter of Organic Growers of Michigan, Coleman, MI.

7. "Management guide for low-input sustainable apple production." 1990. U.S.D.A. Northeast LISA Apple Production Project. Agreement #88-COOP-1-3524.

8. Michigan Legislative Service Bureau Staff. 1992. "Regulation No. 637: Pesticide use." Michigan Admin. Code Issue No. 3. of the 1993 Michigan Register, E. Lansing, MI.

9. Organic Crop Improvement Association Administrative Council. 1993. "Certification Standards." Organic Crop Improvement Association, Parma, MI.

10. Renner, K.A. and J.J. Kells. 1993. "Weed control guide for field crops." Extension Bulletin E-434. Michigan State University, E. Lansing, MI.

EXHIBIT 1

Elizabeth Bergson's apple orchard expenses.

Average yield 400 bu./acre, 15% table-quality, 70% cider, and 15% apple butter.

	Unit Cost
Expenses – Apple production ($/bu.)	
Labor	$ 1.00
Taxes/Insurance	0.90
Inputs (fertilizers, pest control)	1.50
Delivery/Trucking	0.60
Total Cost Per Bushel	$ 4.00
Expenses – Cider (1 bu. = 3.5 gal.)	
300 gallons (including jugs, labor, machinery, transportation)	$50.00
Expenses – Apple Butter (1.3 bu. = 1 case) 12-15 oz. jars	
1 case (including jars, labels, labor, spices, etc.)	$5.00

EXHIBIT 2

ORGANIC GROWERS OF MICHIGAN

CERTIFICATION PROCEDURES AND STANDARDS

I. Certification Procedures

   A. Membership in OGM Chapter

      1. Member information form in & dues paid

   B. Complete and Accurate Farm Records

      1. Field boundaries, aerial map
      2. Crops
      3. Soil amendments
      4. Soil tests
      5. Other information as required by committee

   C. Certification Questionnaire

      1. Completed and signed

   D. Certification Fees

      1. Actual fee set by chapter
      2. Fee may include or be increased by inspector's expenses

   E. Farm Visit

      1. Visual inspection of farm after review of submitted material
      2. Written report of visit

   F. Committee Review and Recommendation

      1. Certified organic status granted
      2. Certified transitional status granted
      3. Status pending/further information required
      4. Certification denied

   G. Certification Renewals

For additional information:

EXHIBIT 3

ORGANIC CROP IMPROVEMENT ASSOCIATION, INTERNATIONAL

1993 CERTIFICATION STANDARDS

PREAMBLE

The following constitute OCIA minimum standards and allowed materials (see Allowed Materials List) for organic certification, and must be met or exceeded by all members of chapters seeking to use the OCIA trademark. All members at large must operate under these standards in order to be certified. Standard proposals and amendments pertaining to agricultural crop production are subject to review at the OCIA Administrative Council meeting which is held in the last quarter of each year. Standard proposals and amendments pertaining to animal husbandry and food processing are subject to review at both the OCIA Administrative Council meeting which is held in the last quarter of each year and the Administrative Council meeting which is held in conjunction with the Annual meeting the first quarter of each year. Proposed changes in standards must be submitted by any member including chapters, farmers at large or corporate members to the Standards Committee at least 60 days prior to the Administrative Council meeting. The Standards Committee shall mail a copy of proposed deletions or additions to the last recorded address of each OCIA Associate at least 45 days before the Administrative Council meets to consider their changes as referred to in the by-laws.

ADMISSIBILITY

1. Certification may be on a whole farm or on a field by field basis. If the latter, all fields of the farm unit must be committed to an ongoing program of organic crop improvement. This program of organic crop improvement must be submitted in writing to the certification committee by the third year of certification and be designed to bring 100% of the farm into transition within at least 5 years following the first certification of any portion of that farm. Exceptions may be granted for portions of the farm which are rented or not under the complete control of the grower or for unexpected and extreme circumstances. Fields may not be rotated in and out of organic production and remain certified.

2. No crop can be sold as "OCIA Certified Organic" if the same crop is also produced elsewhere on the farm using methods or materials that do not conform to these standards, unless the farmer can clearly demonstrate that there exist both the physical facilities and the organizational ability to ensure that there is no possibility of crop mixing. this criterion applies equally to situations when uncertified crop is produced by the same farmer on another farm unit, or is purchased for resale.

3. A field can be certified organic if there has been no use of non-acceptable

materials (insecticide, herbicide, fungicide, fertilizers, etc.) or methods during the three years before harvest.

4. In cases where an adjoining farm is growing heavily sprayed crops, or there is other possibility of contamination, there must exist adequate physical barriers or a 25 foot (8 meter) minimum distance between organic crops and sprayed crops to maintain the integrity of certified fields. When contamination is suspected, the certification agent shall require residue testing. Residue levels must not exceed the lowest legal organic standards existing in any country where the OCIA seal is used.

5. Complete information describing at least three (preferably five) most recent years' production methods and materials, as well as information about current production practices, must be provided. The applicant for certification must also furnish an outline of farm management strategies directed at achieving strict compliance with these standards.

6. To be certified, a farm or field must be managed in accordance with the following required practices listed below, using authorized methods and materials.

REQUIRED PRACTICES

1. Development and implementation of a conscientious soil building program designed to enhance organic matter and encourage optimum soil health.

2. Rotation of non-perennial crops in accordance with accepted regional organic practices.

3. Use of careful management, resistant varieties, intercropping, and maintenance of soil health as the first line of defense against weeds, pests, and diseases.

4. Generation of an audit trail which will permit tracing the sources and amounts of all off-farm inputs, date and place of harvest, and all steps between harvest and sale to the wholesaler, retailer, or final consumer. Certification agents shall recommend denial of certification for inadequate audit trailing.

5. Maintenance of machinery and equipment in good enough condition to avoid contamination of soil or crops with hydraulic fluid, fuel, oil, etc.

6. Use of pre- and post-harvest handling procedures and packaging materials which ensure maximum product quality (appearance, hygiene, freshness, and nutrition) using techniques and materials that are consistent with these standards. Irradiation of certified foods is prohibited.

7. Beginning March 1991, soil testing will not be mandatory for inspection and/or certification purposes. However, OCIA maintains that all members manage their soils responsibly, with the intent to improve soil fertility and tilth through proper management practices. If any problems arise that are associated with nutritionally imbalanced soils, such as poor plant

Appendix A: Regulation No. 637 Pesticide Use

R 285.637.8   DEPARTMENT OF AGRICULTURE   10

(d) So as to cause or allow burying in a land site in a manner that is not in compliance with applicable state and federal solid waste regulations.

(e) So as to cause or allow the storage of pesticides or pesticide-containing materials, including rinsate or wash water, in underground tanks. This prohibition does not apply to watertight catch basins that are used for temporary collection or other recirculating systems as approved by the director.
History: 1992 MR 10, Eff. Oct. 30, 1992.

R 285.637.9 Personal protective equipment.
Rule 9. (1) A pesticide applicator shall follow label directions regarding personal protective equipment.

(2) Personal protective equipment shall be appropriate relative to the potential exposure present at the application site or the type of pesticide use operation to be performed.

(3) Commercial applicators who use a pesticide shall comply with all of the following minimum protective equipment requirements, unless otherwise directed by the pesticide product label:

(a) Long pants shall be worn.

(b) Protective footwear that provides protection from exposure to the pesticide being used shall be worn.

(c) Long-sleeve clothing shall be worn. Short-sleeve clothing may be worn if wash water or waterless soap is immediately available and it is not prohibited by the pesticide label.

(d) Gloves that are impervious to the pesticide in use shall be worn in any situation where the individual's hands are likely to come into contact with a pesticide, unless a program is in place that offers comparable applicator protection.

(e) A person who uses pesticides shall use additional appropriate protective equipment where the likelihood of pesticide exposure exists.
History: 1992 MR 10, Eff. Oct. 30, 1992.

R 285.637.10 Off-target pesticide drift.
Rule 10. (1) Pesticide applications shall be made in a manner that minimizes off-target drift, unless prior authorization and consent as specified in subrule (3) of this rule is obtained from the owner or resident of the land onto which drift may occur.

(2) Before making a pesticide application, an applicator shall do both of the following:

(a) Determine the likelihood of off-target drift.

(b) Determine the direction of possible off-target drift and any sensitive areas that may be impacted.

(3) When pesticide off-target drift is anticipated due to the nature of the application, a drift management plan shall be utilized by the applicator to minimize the occurrence and adverse effects of off-target drift. The plan shall include provisions to secure the informed consent of residents in the affected area before making the application. If, in the course of making an application when off-target drift is not anticipated, there arises an occurrence of off-target drift, the applicator shall notify the residents in the affected area either verbally or with appropriate signs before leaving the application site. The drift management plan shall include drift minimization practices. Such practices may include any of the following:

11   REGULATION NO. 637. PESTICIDE USE   R 285.637.11

(a) The use of the largest spray droplets that are created by a combination of special nozzles, pressures, and particulating agents to accomplish the objectives of the applications.

(b) The use of specialized equipment that is designed to minimize off-target drift.

(c) The use of the closest possible spray release to the target.

(d) The use of the lowest effective rates of application of the pesticide.

(e) The establishment of a no-spray buffer zone. The buffer zone may be treated with nonpowered equipment.

(f) The identification of the maximum wind speed and direction under which applications can be made.

(g) The use of wind shields or windbreaks to contain spray drift or deflect spray drift away from sensitive areas.

(h) Other specific measures stated in the plan that are effective in minimizing the incidence of off-target drift.

(4) Drift management plans shall be in writing. The plan will state the measures to be used and how those measures will reduce the impact of off-target drift. The drift management plan shall be annually reviewed by the person who utilizes the plan.

(5) A record of the sites where the drift management plan was implemented and a copy of the drift management plan shall be retained for a period of 1 year for general use pesticides and 3 years for restricted use pesticides and shall be made available to the director upon request.

(6) Operating under a drift management plan does not exempt an applicator from complying with appropriate federal or state statutes and regulations. However, the department shall consider the presence and use of a drift management plan as a factor in determining appropriate enforcement action.

(7) Pesticide off-target drift shall not include the off-target movement of a pesticide by means of erosion, volatilization, or windblown soil particles after the application of a pesticide.
History: 1992 MR 10, Eff. Oct 30, 1992.

R 285.637.11 Notification and posting requirements.
Rule 11. (1) A commercial applicator shall comply with this rule when making a broadcast, foliar, or space application of pesticides.

(2) When making a broadcast, foliar, or space application of pesticides to an ornamental or turf site, other than a golf course or farm production operation, an applicator shall comply with all of the following provisions:

(a) In addition to requirements specified in R 285.637.12(1) and (2), an applicator shall inform a customer that lawn markers should remain posted for 24 hours, after which time the customer should remove the lawn markers.

(b) Immediately following the application, a commercial applicator shall place a lawn marker sign at the primary point or points of entry. Lawn markers shall be in compliance with all of the following specifications:

(i) Be a minimum of 4 inches high by 5 inches wide.

(ii) Be constructed of sturdy, weather-resistant material.

Appendices

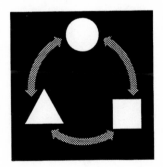

APPENDIX A1:

Abstracts from the Case Showcase

1. Grab-bag Approach to Co-learning: Synthesis, Participation, and Interactive Strategies to Design and Analyze Agroforestry Systems

R.D. William*[1], S. Seiter, S. Cordray, M.L. Rousch, D. Hibbs, and S. Sharrow, Oregon State University

Synthesis learning involves integrating parts into a system, exploring alternatives and examining possible consequences using analysis. In an agroforestry design and practice course, an interdisciplinary faculty introduced students in four sequential weeks to the tremendous complexities of people, animals (wanted and extras), plants (trees, shrubs, and forbs), and societal issues (zoning, taxation, neighbors, etc.). A grab-bag approach was invented by faculty as a fun way for students to grapple with complexity and co-learning within groups. The student's task involved integrating these components into a case or story which increased in complexity each week as a new component was added. Short lectures introduced basic concepts each week and provided a common framework among students from diverse disciplines. These activities contributed to students designing an agroforestry system of their choice during the remaining five weeks of the course.

2. After the Fire

Scott M. Swinton* and Sherrill B. Nott, Michigan State University

This decision case describes a dairy farm partnership faced with what to do after its cow barn and milking parlor burn down. The case is designed for use by undergraduate students with interests in agribusiness management. The immediate focus of the case is on whether to rebuild the dairy facility, what to rebuild, and how to do it. Such questions can be used to explore and apply techniques for investment analysis and strategic planning. The personality conflicts evident in the case raise other issues concerning conflict resolution and the appropriate form of business organization.

3. A Case-based Capstone Course in Crop Management

Steve Simmons*, David Davis, and Gary Malzer, University of Minnesota

A senior-level course has been taught at the University of Minnesota since 1990, which features an integrated approach across the disciplines of agronomy, horticulture, soil science and pest management. It uses decision cases entirely to engage students in the subject matter and achieve course goals. Course objectives include: 1) improving students' competence and confidence in problem identification and in using sound, analytical approaches to problem solving, and 2) enhancing students' abilities to exercise judgement and assess options in crop, soil and pest management. Students respond to the cases individually and in groups using both oral (video) and written approaches. Whole-class discussions are also conducted for each case, as well as interviews conducted with the decision makers featured in some of the cases. Student evaluations of the course have been quite favorable. We conclude that case-based capstone courses are a powerful tool for enhancing students' professional competence.

4. Turning the Tables: Student-developed Decision Cases

Kim Mason* and Steve Simmons, University of Minnesota

Cases have been traditionally researched and developed by faculty and graduate students when used in undergraduate courses. Since 1993, undergraduate students have been given an assignment to develop cases as a term project in an undergraduate capstone course in integrated crop management at the University of Minnesota. The

---

[1]Asterisk denotes author presenting at the Case Showcase.

assignment serves to extend the positive learning experiences gained from the case research process to students. The students make intermediate reports on the progress and process of their research throughout the term. The experience culminates in an extensive oral presentation of the case particulars to their peers. Written outlines of the case and the associated interpretive note are also required. The student-developed case assignment has proven to be very successful for enriching understanding of agricultural dilemmas while creating promising new case topics for further development. Some students continue development of their cases beyond the class and a few have pursued publication of their cases.

5. Structured Discussion In Family or Group Goal Identification

By Paul H. Gessaman*, University of Nebraska-Lincoln

Gessaman, Paul H. "Goals for Farming and Family Life: The Clam That 'Steamed' People." 1995.

_____. "Goals for Farming and Family Life: Whither Leads the Cowpath?" 1995.

_____. "Goals for Farming and Family Life: What Are Sam's Priorities?" 1995.

This set of short vignettes provides stimulus for goal identification and priority setting by family and business unit groups participating in Extension workshops on farm transition planning. Each vignette illustrates issues implicit to the stage of goal identification. The vignette characters deal with issues frequently present in the lives of participants. The vignettes and related discussion questions stimulate participants' thinking and provide easy entre to discussion. Discussion questions highlight issues that often are too sensitive to address in routine discussion within the family or business unit group. In the vignettes and discussion questions, participants find starting points for consideration of similar issues in their own lives. Following participant discussion of the vignette, first steps with the workshop materials are completed without discussion. During individual activities following reading and discussion of each vignette, assembly of candid perspectives is encouraged through physical separation of participants in each

family and/or business unit. When shared, these perspectives receive more nearly equal standing than typically is given to verbal statements. Workshop staff do not prescribe outcomes — in large part, decisions are based on learning through self-discovery. Similar materials and approaches are used in ethical leadership workshops, conflict resolution consultations, community goals workshops, and other learning situations.

6. Consensus on Timber Harvesting in Minnesota

Don MacKay*, Paul Ellefson, and Jon Kilgore, University of Minnesota

This case study focuses on the decisions faced by a staff member of the Minnesota Department of Natural Resources in 1994. The staff member was asked to generate a broad consensus on specific strategies to implement the findings of a controversial, statewide environmental impact statement on timber harvesting. The case study documents the history and development of the environmental impact statement and describes the many stakeholders and issues that would be involved in implementation.

7. Thinking and Learning Dairy Management Systems from Decision Cases

B.J. Conlin*, Jeff Reneau, John Chastain, Jerry Olson, Gerald Steuernagel, and Earl Fuller, University of Minnesota

Thirteen decision cases have been developed for learning dairy management decision making. The case topics cover a variety of dairy farm management issues ranging from narrow topics such as a mastitis problem to whole-farm issues that include a range of problems from family relationships to technical science and farm profitability. Building critical thinking and problem-solving skills has been central to their use. The cases are designed to integrate disciplinary knowledge with a problem-based learning and planning focus. The cases are used as a vehicle to learn to use various dairy farm problem-solving and planning tools, and to introduce specific technical knowledge. Students have included campus-based undergraduate and graduate animal science and veterinary medicine students. Decision cases were first used in an Extension program series for dairy professionals ranging from bankers to feed representatives to veterinarians.

8. Pesticides and the Baby Food Industry: A Case Study

R.C. Herner*, M.A. Uebersax, E. Kabelka, and R. Mazzucchelli, Michigan State University

"One Size Does Not Fit All" was developed for use in a capstone course in the College of Agriculture and Natural Resources at Michigan State University. The case study is based on the recent National Research Council report "Pesticides in the Diets of Infants and Children" and the perceptions of the public concerning pesticides, and their impact on baby food and children. This poster outlines how this case study was developed, the support material supplied to the students, the experiences and observations that we have made using this case study, and how it could be modified to address other questions related to pesticides and food safety in a classroom setting.

9. The Development and Use of Rigorous Whole-farm Planning Decision Cases

Doris E. Mold* and William E. Marsh, University of Minnesota

We have developed a number of whole-farm planning decision cases from actual farms for use in the education of professional, graduate and upper-division undergraduate students, agricultural professionals and livestock producers. The cases are rigorous in that they require the student to synthesize a number of complex components and make decisions at several points throughout the case. The cases are used in group settings where the group must work cooperatively in the role of the livestock producer or advisor to assimilate and distill all of the information given and, in turn, develop a reasonable response to the case.

10. When the Cows Come Home

Cheryl J. Wachenheim*, Illinois State University; Scott M. Swinton, Michigan State University

"When the Cows Come Home" is a case study in market planning that offers an opportunity to apply partial budgeting analysis. Bob Creighton, owner and operator of a 7,000 head feedlot in central Michigan, contemplates the results of a Michigan State University research report that highlights seasonal marketing of cattle as an alternative for Michigan cattle producers. Students assume the role of Mr. Creighton as they consider the immediate and long-run impacts of a seasonal marketing plan on the operation and profitability of the feedlot. The decision making process is complicated by the availability of a three-year contract specifying minimum number of slaughter-weight cattle that must be marketed during the fall months, and the use of seasonally available byproduct feeds complicate. Partial and enterprise budgeting, sensitivity analysis, imperfect information, the nature and use of price data, employee relations, and feedlot management must be considered by students as they work through the case.

Detailed instructor's notes define terminology, review appropriate analytical tools, discuss alternative marketing and production strategies, and provide blank and completed student worksheet templates. "When the Cows Come Home" is most appropriate for use in intermediate and upper level farm and feedlot management and livestock marketing courses.

11. Multimedia as a Complementary Learning Tool to the Case Study Method

Cheryl J. Wachenheim*, Illinois State University; Scott M. Swinton, Michigan State University

Multimedia can be used to enhance teaching by the case study method. Multimedia is, by definition, simply combining more than one form of presentation mode, but most often involves the use of computer-generated visuals. In the classroom, multimedia is an exciting means by which to combine clarity of presentation, visualization of concepts, and use of current, student generated information.

A multimedia package appropriate for use in the classroom and for independent student learning has been designed and is currently being developed for "When the Cows Come Home," a cattle marketing case study. Presentation and discussion of the case are enhanced by a video clip of the calf preconditioning process, a computer-simulated tour of the case farm, and an exercise in which students role play an order buyer by participating in a feeder calf auction. Concurrent presentation of text and spreadsheets provides students immediate feedback on the impact of their assumptions and decisions and serves as a visual scorecard of student input as they "solve the case" in the classroom.

As an independent student learning tool used in the absence of the instructor, multimedia can provide background and review material relevant to a case. It can also help students become more comfortable with the case study method by providing a framework for analysis and immediate feedback. With this tool, the instructor can provide selected information and guidance, as well as monitor student progress and understanding of the material.

12. Projects from the Center for Case Studies in Education at Pace University

William Welty* and Rita Silverman, Pace University

A series of faculty development cases and teaching notes have been designed to encourage faculty to discuss higher education teaching issues relating to diversity issues. Another series of teacher education cases published by McGraw-Hill Primis have been designed for use in undergraduate and graduate teacher education courses. Primis is a new electronic publishing venture which allows faculty to build their own casebooks from a catalog of more than 60 of the center's cases. Materials will also be displayed from an annual summer conference held in cooperation with American Association of Higher Education in which cases, narrative, classroom assessment, and portfolios are used to stimulate reflective practice among university faculty.

13. USDA Challenge Grant at Michigan State University

Eunice Foster*, Michigan State University

In 1993, the U.S. Department of Agriculture project involved 50 Michigan State University faculty members from 14 departments in the College of Agriculture and Natural Resources in a two-day workshop for interactive sessions on learning how to write decision cases. In the summer of 1994, decision case writers and other interested faculty participated in a two-day workshop which focused on the effective use of decision cases in teaching programs. The project will culminate (1) with the publication of cases developed by this project, and (2) with a national conference focused on decision case development and teaching and which will result in the publication of a conference proceedings to be distributed nationally.

14. Environmental Food Safety Decision Cases in Extension Education

By Susan Shumaker, Roger Becker, and Marla Reicks*, University of Minnesota

A series of five decision cases was developed for use in interdisciplinary Extension education. The cases address environmental food safety concerns including the use of biotechnology in food production, herbicides in drinking water, lead poisoning, organic product certification and herbal supplements. The various decision makers in the cases present producer, consumer, and regulatory perspectives. The case experience requires that participants consider principles of risk assessment and management. USDA Extension funded the development of the cases that were presented in the showcase. They are currently being tested by Extension educators in Minnesota prior to dissemination. Each case consists of a one or two page narrative, one or two exhibits and a teaching note. Educators will use the cases in 30 to 60 minute discussion sessions.

15. High Nitrate Showdown at Clear Lake: A Decision Case on Water Quality in an Agricultural Community, Cases A and B

David W. Davis*, University of Minnesota; Melvin J. Stanford, Mankato State University

Water nitrate level in the well water in a small rural town exceeds EPA limits and the town council must decide how to provide safer water 1) on an immediate basis, and 2) on a long-term basis. It must also design a long-term wellhead protection plan. Neighboring potato and corn farmers on the highly porous, irrigated sandy soil around the town are heavy users of nitrogen fertilizers. Solutions to the problem promise to be costly and the town has a very small revenue base. The case is intended for use by university students, county agricultural Extension leaders, community leaders, and by people in various fields who are interested in the multi-sided issues inherent in public policy development.

16. Harvey's Wake-up Call

Josef M. Broder* and Fred C. White, University of Georgia

This case provides a forum for discussion of classroom management techniques in response to changing classroom environments. A distinguished teacher returns to the classroom after spending several years in administration. He discovers that students have changed and that many of his classroom management techniques are no longer effective. The case examines classroom management techniques, teaching philosophies, and student contracts, rewards, and discipline.

APPENDIX A2:

Case Plans Written in the "Writing and Researching a Decision Case" Session

Title: RICE/COTTON DILEMMA OF JOHN CALVIN

Author: Mohammad Jalaluddin

Decision maker: John is a rice farmer of 2,000 acres; 55 years old, has three children; high school graduate of Pine Bluff, Arkansas.

Background: John inherited the farm from his father who traditionally grew rice. They use only one-half of an acre for kitchen garden.

Issues: Rice production cost is high; irrigation water depleted; rice production not profitable. Cotton is profitable, but his soil is not suitable for cotton.

Objectives: Stay on farm; make profit.

Alternatives: Grow certain rice variety that uses less irrigation; lower production costs; increase yields; find a cotton production method or variety to suit his clay soil farm; or do something else.

Target Completion: March, 1996

Title: HOW TO SELL THE HOGS?

Author: Cheryl Wachenheim

Decision maker: Hog farmer (Henry Byck) with 200 sows farrow-to-finish operation. Family with two children, ages 13 & 15 and another on the way. Markets 4,000 hogs a year.

Background: Prices low; lots of concentration (e.g. packers increasingly vertically integrated with producers.)

Issues: Should I join a hog marketing cooperative? More generally, how should I market my hogs in the future?

Objectives: Get highest price for hogs at lowest cost (profit maxinization); flexibility in type of hogs, marketed and timing of sales.

Alternatives: Market on carcass merit basis with order buyer/packer; sell through hog marketing cooperative; look for "window" of best prices; forward contract.

Target Completion: December 31, 1995

Title: A MODERN SIX-HORSE DAIRY FARM

Author: William Heald

Decision maker: Levis Eshleman, dairy farmer with 75 cows in stall barn; has mechanical power, but not electrical. Uses bST, consultants and most modern practices.

Background: Two brothers in business. One is better "doing," the other is a planner, thinker, experimenter.

Issues:	To split into two businesses and borrow money or continue as one business with common business goal but very different personal goals and lifestyles for partners.
Objectives:	Turn over the farm to his eldest son in 15 years and have enough equity to retire.
Alternatives:	Continue farming with brother; split farm equity into two businesses. Start entirely new, non-farm business venture through process of evolution.
Target Completion:	Not given

Title:	**GOING INTERNATIONAL?**
Author:	Art Parker
Decision maker:	Alex Jensen, President of First Quality Avocado Cooperative.
Background:	Avocado Cooperative considering expanding international sales.
Issues:	Should they seek additional sources of supply? Should they cease being a cooperative to get additional supply sources? Should they expand their international sales?
Objectives:	Profit maximization; increase reach of cooperatives.
Alternatives:	Stay as they are, using basically their own members' fruit. Cease being a cooperative.
Target Completion:	Early 1996

Title:	**EXPANDING THE CITY WATER SUPPLY**
Author	Robert Vertrees
Decision maker:	Chief, Division of Water, city of Columbus, Ohio.
Background:	Division supplies water to entire city.
Issues:	Continue to develop the south Columbus well field expansion? (Some landowners have strenuously objected to the initial acquisition and leasing of land to expand the well field; graffiti on barns of farms that have leased; those who have leased have allegedly been forced to move. Well field is in Fairfield County, just south of Franklin County, that includes Columbus.)
Objectives:	Meet the city's projected water supply needs to the year 2010 without a lot of adverse publicity and jeopardization of other supply alternatives.
Alternatives:	Continue the well field expansion plan as already set forth; stop expansion and development of the well field; try to get the Fairfield County residents to get involved in a rational, representative planning process and reach a compromise plan.
Target Completion:	Not given

Title:	**SUSTAINABLE PRACTICES ALONG NATCHEZ TRACE PARKWAY**
Author:	Lynn Reinschmidt
Decision maker:	Gary Mason, superintendent of Natchez Trace Parkway.
Background:	Parkway traces an old Indian trail covering 450 miles across three states (TN, AL and MS). Parkway is federally maintained and includes a narrow strip of land on both sides of parkway.

Issues: Agricultural land along the trace is to be maintained such that the parkway conveys a "traditional agricultural land viewscape."

Objectives: Manage parkway agricultural lands such that agricultural viewscape is maintained while meeting non-point source guidelines on pesticide use.

Alternatives: Lease land; farm the land themselves.

Target Completion: July 31, 1995

Title: ODOR POLLUTION FROM THE NEW PIG FARM

Author: Michael E. Dikeman

Decision maker: Unsure at this point, but probably some state agency (Kansas).

Background: A family farm has been in existence for several generations. A fairly large hog producer moved in less than one-half mile away. The operation causes extensive odor pollution.

Issues: Does the established family farm have a right to live in a reasonably odor-free environment? Does the new hog producer have the right to setup production anywhere? Are there air pollution (odor) standards? Can a state agency make a decision?

Objectives: Make a decision that is legal, based on sound data and information, fair, and able to stand the test of time.

Alternatives: Force the hog producer to shut down production; force him to install extensive manure-handling/air pollution system; do nothing and the affected neighbor will have to accept or move.

Target Completion: May 1, 1996

Title: BECOMING A GOOD NEIGHBOR

Author   Frank Allaire

Decision maker: Managers of an animal production unit, Karl Klutz & Harriet Hostess.

Background: Dairy farm, 300 cows; concerned about improving attractiveness of farm; developing a "sweet smelling" place.

Issues: Balance extra cost with potential to sustain profit.

Objectives: Develop, change animal production operation so more environmentally-friendly while preserving profit.

Alternatives: List possible changes and costs (short, long).

Target Completion: August, 1996

Title: THE CARROT AND THE GUN (THE DEER BAITING ISSUE)

Author   Bob Herner

Decision maker: Director of the Michigan Department of Natural Resources (MDNR).

Background: MDNR's role is to monitor and regulate natural resources in the State of Michigan, including hunting and fishing regulations.

Issues: Should the MDNR prohibit baiting for deer? Is baiting ethical? Is there an anti-hunting/animal rights guide? How does banning baiting affect the supplies of deer food?

Objectives: To make a decision concerning this issue.
Alternatives: Ban baiting completely; ban baiting only on public land; status quo.
Target Completion: Fall, 1995

Title: **THE BLACK HOLE**
Author: Mary Miller
Decision maker: Lynn Connley, irrigation engineer.
Background: Agricultural engineering department at land-grant university has large government-supported grant to develop and build irrigation research project in cotton growing area in southeastern U.S.
Issues: Failure of 20-acre irrigation pond to hold water; competency of survey company (geologists); disappointed politicians; federal funding already extensive; disappointed growers; lack of accountability and credibility of research scientists and engineers).
Objectives: Counteract the sink hole beneath the pond.
Alternatives: Relocate the pond; line the pond; find and plug the hole.
Target Completion: Not given

Title: **LEAFY SPURGE INVADES ANOTHER SOUTH DAKOTA RANCH**
Author: Scott Kronberg
Decision maker: Edward and Judy.
Background: Cattle producers in northeastern South Dakota.
Issues: Much of their land is infested by the weed leafy spurge and they are trying to find better ways to deal with this introduced plant.
Objectives: Stop the spread of this plant on their ranch; eradicate or at least control the plant.
Alternatives: Use insect biocontrol; use herbicides; use livestock biocontrol.
Target Completion: Fall, 1995

Title: **LAND APPLICATION OF BRESOLIDS IN THE RURAL/URBAN INTERFACE**
Author: Gary Pierfynski
Decision maker: Unknown at this time; probably an official with the city of Wichita, KS.
Background: City has expanded its wastewater treatment facilities and must decide on a means for disposing of sewage sludge. Land application in the rural/urban zone is the cheapest alternative.
Issues: Should land application be used despite public opposition?
Objectives: Select the lowest-cost, acceptable, sludge disposal option.
Alternatives: Choose a higher-cost alternative and justify this to the residents of the city.
Target Completion: November 1, 1995

Title:	**[UNTITLED]**
Author:	Nicholas Jordan
Decision maker:	A fictional Chinese couple in a semi-rural area in southern China on the edge of an industrializing city.
Background:	Various negative incentives regarding more than one child; future demographic changes in agricultural structure, population size; rapid changes in Chinese economy; lack of Western social security provisions.
Issues:	Whether to have a second child; have one daughter; need to weigh cost of second child versus future benefits of (especially) a son.
Objectives:	Their own welfare in old age; desire for son based on Chinese traditions; near-term welfare of family.
Alternatives:	Have second child; conceive second child and abort if child appears to be female via ultrasound examination; remain a one-child family.
Target Completion:	June 1997

Title:	**SKIDDING ON THE CLEARWATER**
Author:	Colette DePhelps
Decision maker:	Carl Bjork, forester for less than five years with a private timber company in North Idaho; returned from the University of Washington with mixed feelings regarding conservation vs. preservation and extreme environmental groups.
Background:	Timber company has bad reputation for environmental degradation; attempting to improve image through environmental forestry management IPR program.
Issues:	Whether to cut in a sensitive area near the north fork of the Clearwater River surrounding a high-use hiking trail. EarthFirst! has actively protested timber harvest in this area; company manager says to cut.
Objectives:	Environmental damage likely to occur; manage forest lands for the economic benefit of the private timber company while protecting environmental integrity of private land.
Alternatives:	Cut as ordered by company manager; not cut and ignore order of superior; to cut or not to cut?
Target Completion:	October 1, 1995

Title:	**[UNTITLED]**
Author:	Jack Stang
Decision maker:	VonLubken family: Fritz (father), Joanne (mother), Eric (son); primary decision maker is Fritz.
Background:	VonLubken family orchards, a moderately sized (about 50 acres) Hood River pear orchard; in the family since 1910; produces pears (primarily).
Issues:	Eric and family have returned to orchard after Eric graduated from Oregon State University; expects/wants to take over family orchard. However, his father is not ready to retire, his grandfather is still alive and the size and production system used does not generate the income to support the additional family. What to do?
Objectives:	Generate additional income from orchard operation— enough to support Eric and family.

Alternatives: Eric does not return to family orchards. Increase size of orchard (constraint is that orchard is surrounded by a golf course—another good case); change to high density system (e.g., apples) to generate more income per acre; other possibilities to generate income.

Target Completion: Not given

Title: TREE NITROGEN STATUS AND WATER POLLUTION

Author: Steven Weinbaum

Decision maker: Ken Harbour, age 50, almond grower in Ceres, CA, the San Joaquin Valley; 100+ acres of Hanford sandy loam.

Background: Income based on almond yields; applies 300+ lb. nitrogen (N) per year; irrigation water contains 45 ppm. nitrate; almond is a relatively high N demand crop (yield of 2,000-3,000 lb/acre at 3.2 percent N).

Issues: To apply high amounts of N to insure high yields, balanced against the likelihood of groundwater contamination due to nitrate leaching.

Objectives: To maintain trees in good condition to maximize profitability (yield) while reducing the likelihood of nitrate leaching associated with excessive N rates.

Alternatives: To use diagnostic criteria (leaf analysis, etc.) and objective (quantitative) data on N application and removal in almond crop.

Target Completion: Not given

Title: GONE WITH THE WATER

Author: Dwane G. Miller

Decision maker: A farmer (yet to be identified).

Background: Hilltop farming in eastern Washington; productive rainfed land; highly erodible landscape.

Issues: Tillage systems used in agronomic crop production; conventional vs. minimum till vs. no till.

Objectives: To make money and use environmentally-sound practices for soil conservation; to meet federal policy on residue requirements.

Alternatives: Determine which tillage system best saves soil, maintains yield, gives an economic return and sustains productivity.

Target Completion: January, 1997

Title: [UNTITLED]

Author: Marie Bradhagen

Decision maker: Tom Shephard is a 52-year-old farmer in central Minnesota. He has been experiencing health problems for the past 10 years. Last year he decided to sell his small herd of 50 dairy cattle. Now he is in limbo, too young to retire, but without the training or experience to do another job.

Background: The Shephard farm has been in the family for 150 years; selling the land is not an option; although crops had been used primarily for sustaining the dairy herd, Tom is now considering cash cropping or turning the land into a production site for alternative, high-value crops.

Issues: How much money can Tom afford to invest in setting up a new farming venture? How much maintenance is he willing to do with the cash crops? Can his health withstand the labor required? Other concerns: family, quality of life, and security.

Objectives: Keep the land; make money; maintain lifestyle, but lower labor input.

Alternatives: Hire another hand; rent land.

Target Completion: 2010

Title: **WHERE DO I START?**

Author: Doris Mold

Decision maker: Brian Nelson is a dairy herdsman in east central Minnesota. It has been his lifelong dream to dairy farm on his own. Brian is in his mid-twenties and only has about $25,000 in equity. He does not own cattle or dairy equipment. He studied dairy science at the University of Minnesota but didn't complete B.S.

Background: The Minnesota dairy industry is in decline. The state has been losing hundreds of farms a year. Farms are moving south and west. The average age of farmers has been increasing.

Issues: What options does Brian have to get into dairying? Which option is best for him? Which will the bank finance?

Objectives: To get a start in dairy farming or get a start as an independent dairy farmer.

Alternatives: Work into partnership with family member or with someone outside the family; purchase farm independently (and cows/equipment); lease farm and equipment, purchase cows; lease farm, buy equipment and cows; lease farm equipment and cows; continue as a herdsman; get out of dairying altogether.

Target Completion: April, 1996

Title: **A BURNING ISSUE**

Author: Maria Gallo-Meagher

Decision maker: "Jose Ortega," president of Co-op: Texas Sugarcane Growers Organization

Background: All sugarcane growers in south Texas belong to this co-op; depend on co-op to get cane to the local mill and processed; organized regionally; must burn cane to harvest.

Issues: Burning of the cane pre-harvest: Is it environmentally safe? Residents disapprove of the ash that falls on their property because it is messy, may harm their health, pollute the water, etc. Issue is also one of human rights—illegal aliens enter the fields and could be harmed by the burning. So a number of groups (environmental, residential, Mexico) want burning stopped; growers want it to continue.

Objectives: Appease fears; make burning safer and make money for growers.

Alternatives: Stop the burning; continue burning; other ways to harvest sugarcane.

Target Completion: July, 1996

Title:	[UNTITLED]
Author:	Claudia Parliament
Decision maker:	Manager of Seward Food Co-op, recently hired away from a very successful food co-op in northern California.
Background:	Seward is successful neighborhood co-op that has been in existence for more than 20 years. The store specializes in organic and bulk foods. Seward is one of seven local food co-ops in Twin Cities area of Minnesota.
Issues:	Recently, an investor-owned food chain opened in Twin Cities area. The new chain looks like a food co-op offering similar products. This new competitor has forced local co-ops to consider various forms of coalition to reduce costs and more effectively market and compete.
Objectives:	To at least maintain, but preferably increase, member base and goods and services provided.
Alternatives:	Merge with other local food co-ops; do nothing; form some sort of coalition with other food co-ops.
Target Completion:	June, 1996

Title:	A CONTRACT ON HOGS
Author:	Scott Swinton
Decision maker:	John Smith raised hogs farrow-to-finish in southwestern Michigan until 1994. With hog prices at an historic low in 1994, he sold his hogs and quit production.
Background:	Since 1990, the hog industry has been moving rapidly from individual farm production to vertically-integrated contract production, tying producer to slaughterhouse with standardized production practices.
Issues:	In early 1995, John was offered a contract to feed hogs under strict specifications for a nearby corporation. Should he accept?
Objectives:	Financial security with retirement in offing; like to keep farming.
Alternatives:	Contract hog production on a five-year contract (with new facility and attached debt); off-farm job; custom farm for others.
Target Completion:	November 1995 (if decision maker and contracting corporation agree to collaborate).

Title:	PRIVATE USE OF RESEARCH LAND FOR HOUSEHOLD COMMERCIAL FARMING
Author:	Bernard Mtonga
Decision maker:	Director of the Agricultural Research Station; the most senior person (academically), but only 30 years old and supervises older and younger officers.
Background:	The research station has 10,000 acres of research plots. It is situated among surrounding villages. Each family household has some land for private use. Research plots should not be disturbed.
Issues:	Charcoal burning practiced as land-clearing strategy. Should the land be used for private commercial farming? Use any fertilizers? Restrict farming to small plots already assigned?

Objectives: To preserve research land for quality data; to satisfy the workers with an option; to provide sound farming practices to workers.

Alternatives: Restrict farming or release the land? Do nothing about it; ask workers to resign and take up commercial farming; fence the plots.

Target Completion: 1996-97 or ASAP

Title: **CALLING KING SOLOMON**

Author: Eunice Foster

Decision maker: Director, Michigan Department of Agriculture.

Background: PCB has contaminated feed. Cattle are dying, people have reported symptoms of illness. Extent of contamination unknown; highly charged public atmosphere.

Issues: What to do about cattle that may have been contaminated? How to deal with people who've reported symptoms of illness? What to do with contaminated milk? How to prevent further contamination?

Objectives: To prevent contaminated beef and milk from getting into the food chain; allay community fears; decrease tension in the community.

Alternatives: Test cattle; destroy all cattle suspected of contamination; dispose of all milk; hold community meetings.

Target Completion: August 30, 1995

Title: **SQUEAKY CLEAN CORN**

Author: Roger Becker

Decision maker: Carl Corn, farmer; dairy/cash crops on 500 acres, 250 irrigated.

Background: A typical sand-plain farmer, corn, soybeans, alfalfa.

Issues: Groundwater contamination with herbicides; voluntary program to use NRCS approved production plan for SP-53 cost share grant.

Objectives: Profit; long-term weed control; risk management.

Alternatives: More rotations; switch herbicides; cultivate/rotary hoe; cover crops; alternative crops/practices.

Target Completion: February 1, 1996

Title: **[UNTITLED]**

Author: Al Wysocki

Decision maker: Manager of a crop improvement association.

Background: Field seed industry (i.e. dry beans, soybeans, wheat, etc.); structure the crop improvement association according to member interests; particularly focus on public variety field seeds.

Issues: Future of public variety seed vs. privatization; more profitability of the public variety seed industry; future of public breeding programs; proper role of cooperative marketing arrangements among seed producers.

Objectives: To create a system that is profitable for the public variety seed producers.

Alternatives: Do nothing about current situation; organize a cooperative marketing effort (provide tools, but can't actually market as a crop improvement association); develop trading programs for producers to do a better job marketing.

Target Completion: November 30, 1995

Title: THE DILEMMA OF PROSO MILLET IN SWEET CORN

Author: Ray William

Decision maker: Vegetable farmers in Oregon.

Background: Vegetable growers struggle with wild proso millet in sweet corn; weed is spreading fast in Willamette Valley.

Issues: Control practices are not working; weed is being spread around valley.

Objectives: Help students/farmers struggle with management of weed and learn about weed biology and control. What should vegetable growers do?

Alternatives: Establish differential between sweet corn and millet; reduce N response of millet; maintain herbicide at/above soil surface; design post-control practices for knee-high corn; prevent spread and rotate crops with good control options.

Target Completion: When I can use it in weed shortcourse.

Title: FULL SPEED AHEAD AT LAKE POPLAR FARM

Author: David Davis

Decision maker: Edward Cooper, crop consultant; chair, county planning and zoning committee.

Background: Lake Poplar Farm is Cooper's largest client; grow potatoes, dry beans under irrigation and some dryland field crops. Farmer is in suburban area and is a highly visible, high-chemical user. He is clearing trees from suburban space for more land; doesn't follow zoning and PCA rules.

Issues: Right-to-farm; land use; largest client. Who decides if shelter belt trees can be cut? Water flow and source of high nitrate N in groundwater; protocol needed to produce a profitable crop.

Objectives: To stay in business; to be a good citizen; to protect his client.

Alternatives: Drop the client; write a decision case and share with the farmer; withdraw from county zoning responsibility; influence Lake Poplar Farm to modify practices.

Target Completion: February, 1996

Title: HEAVY METAL GOLFERS?

Author: John Graveel

Decision maker: City of South Bend, lawyers, superintendent.

Background: Black Thorn Golf Course.

Issues: Extent of heavy metal contamination of the Black Thorn Golf Course caused by sludge application in the past.

Objectives: To determine whether heavy metal contamination occurred and, if so, its extent; to open the golf course (status quo); to satisfy the public about contamination status.

Alternatives: Close the course (happens to be the best new course in the U.S.); dig into the contaminated soil; keep the pH above 6.0.

Target Completion: 1996

Title: IN LIVING COLOR: INTEGRATED PEST MANAGEMENT PRACTICES IN PUBLIC GARDENS

Author: Anne Hanchek

Decision maker: Peter Olin, director of Minnesota Landscape Arboretum; associate professor, Department of Horticultural Science, University of Minnesota; superb fundraiser and practicing/teaching landscape architect.

Background: Unit of Department of Horticultural Science, University of Minnesota; budget two-thirds dependent on earned income from visitors and donations; mission is research, education, and "to inspire and delight visitors;" barely above-water financially; non-university board of trustees.

Issues: Should the main annual beds be reduced in size or moved? Heavy disease problems in beds require expensive chemical management. Large size takes up much paid staff time and volunteers. Shrubby area is beloved by public and supported by donors.

Objectives: Please public; please staff member in charge who obsesses over garden; please IPM staff who can't handle disease problems; please supervisor who wants effort shifted; please donors; reduce overall operating costs; maintain good looks of Arboretum. Can current chemicals be used in the future? Are they perceived as a public hazard?

Alternatives: Reduce size; retain sizes; move garden; reassign staff; ask for donors' opinions; ask donors for more support; change cultivars; change display to IPM education; let supervisor decide.

Target Completion: Not given.

Title: [UNTITLED]

Author: M. Wiedenhoeft

Decision maker: Maine Department of Environmental Protection; probably a committee or board.

Background: A California company wants to produce wind power. After research, they find that wind speeds in the mountains in western Maine are among the highest in the U.S.

Issues: Wind power can be used to produce electricity, replacing nuclear power. Will birds be killed? About 125 miles of roads have to be built in the mountains in order to set up the windmills. This area has some fragile landscapes.

Objectives: To protect the environment (animals, plants, earth, water); to provide a safe environment for the people of Maine.

Alternatives: Not give permit; give permit without change in plans; reduce the number of windmills which would reduce the number of miles; could fewer miles of road be constructed and still have successful electricity production?

Target Completion: July 1, 1996

Title:	**DO YOU REALLY WANT TO FARM?**
Author:	Monte Vandeveer
Decision maker:	Larry Jackson, 35 years old, single, teaches high school and coaches two sports/year in a small Kansas town; wants to farm full time, but has few resources; currently renting three small farms and farming part time.
Background:	Dryland wheat in Central Kansas. Cow/calf and stocker calves in Kansas; teaching salary scale.
Issues:	Should Larry quit teaching to farm? Can Larry obtain the resources needed to be a successful full time farmer? Does he have the management ability to run the farm business?
Objectives:	Enjoy the farming lifestyle—be own boss, work outdoors, etc.; income and standard of living at a certain level.
Alternatives:	Teach and farm part time; teach only; quit teaching and farm full time (immediate transition over time).
Target Completion:	End of 1996

Title:	**NO, YOU CAN'T GO!**
Author:	Brent Pearce
Decision maker:	Pam, a student getting married and moving the last semester of senior year from Ames, Iowa to Houston, Texas.
Background:	She has taken all her school work at Iowa State up to her last semester; she wants to enter Rice University and transfer 16 credits from Rice to ISU for graduation from ISU; sent proposal to transfer credit to the college.
Issues:	Pam's proposal has been refused by an associate dean at ISU saying that marriage is not a good reason. How can she move to Houston and still graduate from ISU?
Objectives:	To be allowed to use credits at another school to complete graduation; to be effective in changing her academic program; to be taught in a freshman orientation course.
Alternatives:	Remain at ISU; request that the college grant exemption to rule; request advisor's help in getting an exemption.
Target Completion:	January, 1996

Title:	**TO KILL A COCKLEBUR**
Author:	Scott Reuss
Decision maker:	Producer, agricultural grain production.
Background:	Conventional farming practices used on the farm for many years with herbicides always present; reduced tillage options used whenever feasible.
Issues:	Effect of herbicides on environment; ethics of using pesticides; practical benefits of reduced tillage; measure of safety of work environment.
Objectives:	Maintain present practices; have economically viable farming system; be ecologically sound and conserve soil's health.
Alternatives:	Maintain present practices; reduce herbicide input; increase herbicide input and thus reduce tillage.
Target Completion:	No plan to develop at present

Title:	**FAMILY FARM FINANCIAL CRISIS**
Author:	Sally Noll
Decision maker:	Turkey producer on family farm, in early 50's; taken on leadership to initiate an analysis process with other family members in the farm operation to the extent allowed.
Background:	Decision maker is part of family farm organization which includes his father, brother, and potentially one son and a nephew. Farm has several operations: turkey production, turkey breeding, crops, swine farrowing and finishing, feed mill (with outside sales) and grain storage/marketing.
Issues:	Need to reduce debt/equity ratio for family farm to improve ability to get financing. Issues: Who controls family organization (reduce father's involvement after 70 years); business organization; farm and personnel management; financial situation.
Objectives:	Identify how to reduce debt/equity ratio. Improve performance, costs of production in the various operations. What needs to be done to include younger son/nephew into business?
Alternatives:	Reorganize various farm operations into separate identities; sell various parts of the operation; one person in control of whole operation; rent/sell land; reduce number of workers.
Target Completion:	July, 1996

Title:	**RENOVATE OR BUILD NEW**
Author:	Andy Skidmore
Decision maker:	Fred and Jack vander Hooven, brothers and partners in 150-cow dairy.
Background:	Farm is five miles from major, growing urban city. One brother "workhorse," other brother "playboy" and dreamer, and wants to be "big man on campus."
Issues:	Decision to remodel or build new milking parlor; dilemmas: urban encroachment, conflicting goals of partners, profitability.
Objectives:	Profitability improvement.
Alternatives:	Do nothing; build new milking parlor; renovate old milking parlor; split partnership.
Target Completion:	January, 1996

Title:	**HEADSPACE AIR IN RETORT SAUSAGE POUCHES**
Author:	Howard Zhang
Decision maker:	John Brown, plant manager, Cinpac.
Background:	Cinpac is a military ration producer. They produce packaged food product for three-year shelf life at ambient temperature.
Issues:	Headspace air in retort sausage packs causes rapid oxidation of lipids and off-flavor development. Cinpac has no on-line detection. Large batches of product may stay in plant and be trashed when one out of 10 packs contain more than 10 cc headspace air.
Objectives:	Reduce loss of product.
Alternatives:	Stay using conventional quality assessment; install off-line sorter for product batches that failed quality assessment; install on-line sorter to eliminate product loss.
Target Completion:	Not given.

Title: AMANA FOREST MANAGEMENT – SUSTAINABLE OR NOT?

Author: Joe Colletti

Decision maker: Larry Schultz, forester for a large private farming/forestry operation in Midwest (Iowa), B.S. degree, has part-time assistants (summers).

Background: The Amana Society is a religious-based German community with agricultural/forest production activities and tourism. Amana forestry is a "profit center" as is agriculture within the corporate Amana Society structure.

Issues: How to generate net income through timber harvests and sustain recreational use (hiking, hunting) of all forest lands. Quality oak, walnut, hickory and ash have traditionally been produced, but citizens are concerned about environmental impacts.

Objectives: Achieve CEO-determined return on investment of timber harvest (12%); achieve society recreational use objectives (hiking & hunting); sustain forest resources.

Alternatives: Continue the sequence of annual harvests based on traditional timber harvest regulation (short rotation age); apply modern ecosystem-based timber harvest concepts (long rotation age).

Target Completion: January 1, 1997

Title: THE MUDDY WATER (BUFFER STRIP ADOPTION)

Author: Dick Schultz

Decision maker: Don Tisdal, farmer and maintenance engineer at Iowa State; children are grown and gone; uses conventional farming practices; loves hunting and wildlife; has grandchildren.

Background: Eighty-acre family farm with meandering creek through it; corn and soybeans on 85% of area, pasture along creek; uses farm programs to insure income from crops.

Issues: Water contamination from surface sediment stream bank collapse, tile inflow; concern for environment; desire to diversify farm; provide opportunities for grandchildren; stewardship; government control; loss of income.

Objectives: Improve stream environment; establish change at minimum cost (installation and loss of income); increase wildlife.

Alternatives: Remove cattle from along creek and cultivate pasture grasses; do nothing and rent pasture to neighbor with cropping to creek edge; establish a buffer strip of trees, shrubs, native grasses; establish stream bank protection and continue grazing practices.

Target Completion: January 1, 1996

Title: [UNTITLED]

Author: Roger Ruan

Decision maker: Plant superintendent.

Background: Feed manufacture plant.

Issues: Flowability of feed; adding sand or clay (within the limits) will improve flow significantly; adding sand improves profit, but decreases nutritional value.

Objectives: Improve flowability to improve process efficiency and product consistency.

Alternatives: Invest in research to find ways to improve flowability.

Target Completion: Not given.

Title:	GETTING THE TRUCK THROUGH THE DOOR
Author:	Paul Dietmann
Decision maker:	Kevin Kishnan, a dairy producer in Door County, WI, uses intensive rotational grazing and is well on his way to organic certification.
Background:	Kevin farms where his family has since the early 1900s. Organic milk processor is ready to begin picking up milk in Door County later this summer. The processor is in western Wisconsin. Door County, in eastern Wisconsin, is a big tourist area. Kevin believes that a very good opportunity exists to market organic meat and vegetables to wealthy and health-conscious tourists from the Chicago area.
Issues:	The processor will pick up every other day. Kevin's bulk tank only holds one day's production. Some of the farmers in the county are seasonal producers which means that there may not be enough milk to send the truck during the winter.
Objectives:	Kevin wants to continue milking year-round; would like to ship organic products; has no intention of expanding herd; would eventually like to quit milking altogether.
Alternatives:	Continue shipping conventionally, daily market organically by buying a new bulk tank for $20,000 or go seasonal in reverse (milk more cows in winter).
Target Completion:	October, 1995

Title:	THE HORSE HEAVEN HILLS CASE: SHOULD AERIAL APPLICATION OF HERBICIDES BE BANNED?
Author:	Carol Mallory-Smith
Decision maker:	A state department of agriculture.
Background:	Organization is responsible for pesticide use and settling questions of chemical trespass.
Issues:	Herbicide drift from wheat fields to vineyards and residential areas; aerial application of herbicides.
Objectives:	Find solution that protects public without jeopardizing wheat growers; sets an acceptable precedent for future suits.
Alternatives:	Buffer zones; ban or not ban particular herbicides; ban or not ban aerial application.
Target Completion:	Not given.

Title:	THE BURNING QUESTION
Author:	Patricia Lindsey
Decision maker:	Mr. & Mrs. Soyer, Soyer Farms; grass seed growers; own and operate large farm in Willamette Valley, Oregon; includes seed cleaning, bagging, storage facilities.
Background:	Growing grass seed is a major industry in Oregon and produces vast majority of U.S. production of several types. Straw has typically been disposed of through field burning. Field burning is now restricted for environmental reasons.
Issues:	Burning is increasingly restricted. Chemical alternatives for disease/pest control are not as effective; pest resistance may develop; EPA restrictions limit pesticide availability. What to do with straw if not burned; genetic purity, pest risk without burning.

Objectives:	Not lose money; keep farming; not be cited for violations; minimize risks.
Alternatives:	Burn; burn part (some fields, not others); don't burn (store straw, bale straw, mulch, find market, use chemical disease/pest control methods, grow something else).
Target Completion:	Not given.

Title:	**CAN RESISTANCE BE RISKY?**
Author:	Ed Braun, Iowa State University
Decision maker:	Not sure; persons in Florida who must decide whether or not to allow release of genetically-engineered (virus-resistant) cucurbits.
Background:	Asgrow and others have developed virus-resistant cucurbits. Some environmental groups feel they should not be released.
Issues:	Environmental risks associated with genetically-engineered plants—development of new strains of virus; "escape" of the genes into weed populations.
Objectives:	To assess the risks and benefits of releasing genetically-engineered plants.
Alternatives:	Allow or prevent the release. Perhaps decide what safeguards are needed.
Target Completion:	May, 1996

Title:	**FARM MANAGEMENT ABCDEFG**
Author:	Hong Wang
Decision maker:	Al (grandfather), Barry (father), and Chris (son). They own a dairy farm in mid-Michigan. Both Barry and Chris are operators. Barry's wife, Diane, is in charge of bookkeeping. Chris has a B.S. in animal science from Michigan State University. His wife, Ellen, works off the farm. They have an 8-year-old son, Frank and a 6-year-old daughter, Gloria.
Background:	They own 800 acres of tillable land and rent 450 acres to grow corn, soybean, wheat and hay, mostly for feed; 270 cows produce 24,000 lb. milk daily; hire 4 full-time laborers and a herdsman; sell the milk, wheat, soybeans and livestock to a local market. Neighbor farmer hauls manure daily to his farm for a small payment.
Issues:	Farm owned for only 7 years, need to generate more cash to pay the debts on initial investment; extend the dairy enterprise (milk sells well); need a larger milking parlor, but not sure can take extra loan; manure a potential problem because of groundwater pollution, causing concern to environmental group and the town. Neighbor farmer may quit hauling their manure; animal rights concerns requiring more investment in barns and other facilities.
Objectives:	Generate more cash income and expand business; reduce debt/asset ratio; keep business healthy to ensure Al's retirement and avoid risky investment; stay in dairy production to use Chris's specialty; increase scale to compete, in case laws change against them; pass the farm on from generation-to-generation.
Alternatives:	Keep operation same size, but manage the crop enterprise more efficiently to meet the cash flow need; expand dairy production gradually, without investing in a new parlor or new barns; expand dairy production rapidly with investment in new buildings, facilities and manure storage and process equipment, hire more laborers.
Target Completion:	Not given.

Title:	PERFORMANCE – BASED PAY AT XYZ COOP
Author:	Rob King
Decision maker:	Board chair of a local farm supply cooperative for 10 years; is also a farmer.
Background:	Co-op has annual sales of $25 million in seed, fertilizers, pesticides, agricultural services, hardware, tires, convenience store.
Issues:	Should the board develop or adopt a compensation system for the general manager that includes significant bonuses based on pre-determined incentives?
Objectives:	Make manager more responsive to cooperatives; not possible to give manager ownership stake. System should be fair and should not create "perverse" incentives.
Alternatives:	Stay with current system; salary increase with an occasional "pat-on-the-back" bonus; adopt pre-determined incentive system.
Target Completion:	December, 1996?

Title:	"DOGS, STUDENTS AND SMALL COLLEGES" OR "A HEARTFELT MATTER"
Author:	Ted Johnson
Decision maker:	Department chair Jane Roe, ecologist 15 years teaching/research involving pollution at St. George's College.
Background:	Small liberal arts college; research projects with undergraduate students. Main focus is teaching (3,000 students, no graduate students); undergraduate research is encouraged.
Issues:	New faculty member being recruited who has extensive research background; uses dogs in cardiophysiology; has a large, transferable grant. Should his research be encouraged? Appropriateness of research model: dogs in small facility (student and community concern); useful teaching model.
Objectives:	Recruit an outstanding faculty member; maintain departmental harmony and balance "Joe Nixon;" be a good community partner; honor individual faculty rights.
Alternatives:	Adapt facilities to meet faculty member's needs; suggest faculty member use a different (smaller) animal model; use alternative research facility off-campus; not recruit that individual.
Target Completion:	August, 1996

APPENDIX A3:

Agenda for the Workshop

TEACHING AND LEARNING WITH CASES: PROMOTING ACTIVE LEARNING IN AGRICULTURAL, FOOD AND NATURAL RESOURCE EDUCATION

JULY 6-8, 1995
Oak Ridge Conference Center
Chaska, Minnesota

Agenda

Thursday, July 6

10:00 a.m.	**Registration Check-in Begins**	*Lobby*
11:30 - 1:00 p.m.	**Lunch Buffet in the Minnesota Seasons Restaurant** *You must wear your name badge in order to be served.*	*First Level*
1:00	**Conference Welcome and Overview** Anita Nina Azarenko, Oregon State University Eunice Foster, Michigan State University Steve Simmons, University of Minnesota	*Room 208*
1:45	**The Basics of Case Teaching** Josef Broder: "An Honest Face" Kent Crookston: "The Worth of a Sparrow"	*Room 208*
3:15	**Refreshment Break**	
3:30	**Classroom Techniques in Case Teaching ***	
	Group A Facilitator: Steve Simmons	*Room 208*
	Group B Facilitator: William Welty	*Room 200*
	Group C Facilitator: Chris Peterson	*Room 204*
	Group D Facilitator: Oran Hesterman	*Room 216*
5:00	**Case Teaching and Team Assignments ***	
	Teams A- 1,2 &3 Coaches: Steve Simmons Andy Skidmore	*Room 208*
	Teams B- 1,2&3 Coaches: Josef Broder William Welty	*Room 200*
	Teams C- 1,2&3 Coaches: Chris Peterson Jack Stang	*Room 204*
	Teams D- 1,2&3 Coaches: Kent Crookston Oran Hesterman	*Room 216*
6:00	**Dinner Buffet in the Minnesota Seasons Restaurant**	*First Level*
7:30	**Case Teams Prepare for Teaching**	*Team Choice*

* *See blue "Group and Teaching Teams Assignments" sheet in conference folder.*

Friday, July 7

6:30 - 8:00 a.m.	**Breakfast Buffet in the Minnesota Seasons Restaurant** *You must wear name badge in order to be served.*		*First Level*

8:30   **Teams Teach Assigned Cases**

Teams A- 1,2 &3	Coaches:	Steve Simmons Andy Skidmore	*Room 208*
Teams B- 1,2&3	Coaches:	Josef Broder William Welty	*Room 200*
Teams C- 1,2&3	Coaches:	Chris Peterson Jack Stang	*Room 204*
Teams D- 1,2&3	Coaches:	Kent Crookston Oran Hesterman	*Room 216*

10:00   **Refreshment Break**

10:35   **Teams Teach Assigned Cases - continued**

12:00   **Lunch Buffet in the Minnesota Seasons Restaurant**   *First Level*

1:00 p.m. **Teams Teach Assigned Cases - continued**

2:20   **Refreshment Break**

2:45   **The Basics of Designing Cases into a Course**

Group A	Facilitator:	Josef Broder	*Room 208*
Group B	Facilitator:	Chris Peterson	*Room 200*
Group C	Facilitator:	Steve Simmons	*Room 204*
Group D	Facilitator:	Jack Stang	*Room 216*

4:00   **Free Time**

6:00   **Dinner Buffet in the Minnesota Seasons Restaurant**   *First Level*

7:00- 9:00	**Case Showcase**	*Break Area*

Saturday, July 8

7:00 - 8:30 a.m.	**Breakfast Buffet in the Minnesota Seasons Restaurant** *You must wear your name badge in order to be served.*		*First Level*

8:30	**Techniques for Researching and Writing Cases** Josef Broder, Kent Crookston, Emily Hoover	*Room 208*

10:15   **Refreshment Break**

10:30   **Concurrent Sessions I ***

      A. Managing the Case Experience I   *Room 204*
      Student assessment, teaching and learning
      style interaction
      Facilitator: William Welty

      B. Managing the Case Experience II   *Room 208*
      Developing written and oral assignments
      using cases; case enrichment through videos,
      debates, discussion teams, guests, role play;
      student-developed cases
      Facilitators: Steve Simmons, Jack Stang, Kim Mason

      C. More on Developing Course Design   *Room 216*
      Class size, student experience levels, levels
      of course offerings
      Facilitators: David Davis, Chris Peterson

      D. Case Use in Extension   *Room 206*
      Facilitators: Marla Reicks, Ray William

12:00   **Lunch Buffet in the Minnesota Seasons Restaurant**   *First Level*

1:15 p.m. **Creating a Vision for Land Grant Universities**   *Room 208*
      and Their Teaching and Learning Programs
      Tom Warner, W.K. Kellogg Foundation

2:00   **Concurrent Sessions II ***

      A. Managing the Case Experience I   (repeated)   *Room 204*
      Student assessment, teaching and learning
      style interaction
      Facilitator: William Welty

      B. Cooperative Learning   *Room 208*
      Experiential exercises that build students'
      capacities for case learning
      Facilitator: Emily Hoover

      C. Master Session   *Room 206*
      Discussion of case teaching issues for
      experienced teachers
      Facilitators: Josef Broder, Kent Crookston

3:30   **Refreshment Break**

3:45   **Maintaining the Momentum:**   *Room 208*
      Case Teaching for the Long Run
      Anita Nina Azarenko, Eunice Foster, Steve Simmons

4:30   **Evaluation and Adjournment**

    ***   See yellow "Saturday Concurrent Sessions" sheet in conference folder.*

5:00	**Dinner Buffet in the Minnesota Season's Restaurant**	*First Level*
	For participants staying at Oak Ridge Saturday night.	
6:30	**Optional Excursion to Mall of America**	*Lobby*
10:30	**Mall Pick-up and Return to Oak Ridge Conference Center**	
	Meet the bus at the Transit Hub, lower level of East Broadway	

Resource Room

An information table will be staffed throughout the conference in Room 210. In addition display tables will be available for sharing program materials and other resources.

Financial Support

This conference was funded in part by the W.K. Kellogg Foundation. The W.K. Kellogg Foundation was established in 1930 to "help people help themselves." As a private grantmaking organization, it provides seed money to organizations and institutions that have identified problems and designed constructive action programs aimed at solutions.

Additional funding was also received from a USDA Challenge Grant through Michigan State University.

Planning

"Teaching and Learning With Decision Cases" was planned and organized by faculty with expertise in decision case teaching and writing from Michigan State University, Oregon State University and the University of Minnesota.

APPENDIX A4

Workshop Facilitators

Dr. Anita Nina Azarenko is an associate professor of horticulture and is a co-director of the Bioresource Undergraduate Research Program at Oregon State University. She teaches and facilitates learning in an upper level fruit and nut production course and coordinates experiential learning opportunities for undergraduate students interested in research in agricultural, forestry, food and environmental sciences. Anita has developed and used case studies in her advanced pomology class. Her research spans the continuum of different research methods of inquiry from basic, where she focuses on the floral biology of fruit and nut crops, to more systems research where integrated fruit production is the focus.

Dr. Josef M. Broder is a professor of agricultural and applied economics at the University of Georgia-Athens. He serves as undergraduate coordinator for programs in agricultural economics, agribusiness, and environmental economics and management. He received the 1993 National Award for Excellence for Outstanding Teachers in Food and Agricultural Sciences sponsored by the U.S. Department of Agriculture. Josef is best known for his extensive publications on interactive teaching and undergraduate student affairs. He has developed and used case studies for classroom instruction and professional development.

Dr. Kent Crookston is professor and head of the Department of Agronomy and Plant Genetics at the University of Minnesota. He has an extensive record of research, teaching and international work as a crop physiologist. He was elected fellow of both the American Society of Agronomy and the Crop Science Society of America in 1992. Kent was a pioneer in the development and delivery of decision cases within agriculture. In 1991, he co-chaired the first national conference on case education in agriculture. He has used case studies in the classroom, in industry, state organizations and university outreach and has led numerous workshops

to promote decision cases nationally and internationally.

Dr. David W. Davis is professor of horticultural science at the University of Minnesota and works in the area of vegetable agriculture. He teaches commercial vegetable agriculture and is a co-instructor in integrated management of cropping systems. Dave's research emphasizes vegetable breeding, with a focus on crop adaptability, new crops, and disease and insect resistance. He has written and published decision cases and has used decision cases in classroom teaching and in mentoring graduate students.

Dr. Eunice Foster is a professor and crop physiologist in the Department of Crop and Soil Sciences at Michigan State University. Her teaching experience includes self-contained sixth-grade classes in the public school system, crop production in the Institute of Agricultural Technology at MSU, and undergraduate courses in crop science and crop physiology. Eunice strives to actively engage all students in the learning process and to facilitate cooperative learning, critical thinking, and the development of speaking and writing skills. She views decision cases as one method to help achieve these goals. Eunice is principal investigator for MSU's USDA challenge grant on "Decision Case Use in Agricultural Sciences: Researching, Writing and Teaching."

Dr. Oran Hesterman is professor of Crop and Soil Sciences at Michigan State University. He was first exposed to decision case learning as a student in the business college at the University of Minnesota. He currently teaches a class in cropping systems management to undergraduate students and uses the decision case method extensively in this class. He has authored or co-authored seven decision cases dealing with issues in crop production management. Oran's research and Extension emphasis is in forage systems management, with a focus on utilizing legumes as a substitute for nitrogen fertilizer. He also has a background in leadership development, having participated as a Kellogg National Fellow from 1987 to 1990.

Dr. Emily Hoover is an associate professor in the Department of Horticultural Science at the University of Minnesota. Her interest in decision cases began with Project Sunrise, a Kellogg-sponsored grant at Minnesota. She has written two cases and taught with other ones in her horticultural science courses. Her research interests combine the use of expressive writing with cooperative learning in courses. She has been a faculty mentor for new faculty members at the University of Minnesota and has been very active in teaching and curricular revision.

Dr. Christopher Peterson teaches and conducts research in agribusiness management in the Department of Agricultural Economics at Michigan State University. Chris joined the MSU faculty in the fall of 1991 to help lead a new undergraduate curriculum in agribusiness management. He specializes in the strategic management issues facing firms in agricultural input supply, production, commodity assembly, and initial processing. He has been writing and teaching with cases for nearly 15 years, using the method with agricultural technology students, undergraduates, graduate students, and industry professionals.

Dr. Marla Reicks is an assistant professor of nutrition at the University of Minnesota and an Extension nutritionist with the Minnesota Extension Service. Her research and programming interests include nutrition and food safety education for limited-resource audiences. Over the past three years, she has developed and implemented decision cases related to nutrition and food safety/environmental issues for secondary and Extension audiences.

Dr. Steve Simmons is a professor in the Department of Agronomy and Plant Genetics at the University of Minnesota. He has been a leader in his college's effort to adapt decision case education and research for disciplines within food, agriculture, natural resources, and environmental

sciences and serves as chair of the Program for Decision Cases. He has conducted numerous workshops to stimulate understanding of decision cases and their applications to higher education. In 1991, he co-chaired a national conference on case education in agriculture that was attended by academicians from throughout the United States. He has authored decision cases for publication and regularly uses them in his courses. He currently co-teaches an interdepartmental "capstone" course for majors that is formatted entirely around decision cases and has also incorporated decision cases into a new agricultural ecology course within the University's liberal education curriculum.

Dr. Andrew Skidmore is assistant professor of dairy management at Michigan State University. He teaches undergraduate courses in dairy management and advanced enterprise management. He participated in the development of the Animal Management Advancement Project and is currently involved in designing the Human Resource Management and Development Module. He is also a co-investigator on MSU's USDA Challenge Grant on "Decision Case Use in Agricultural Sciences."

Dr. Scott M. Swinton is an assistant professor of agricultural economics at Michigan State University. He teaches undergraduate farm business management and graduate agricultural production economics. Over the past three years, he has experimented with active learning techniques, including farm management decision cases. His research in agricultural production and environmental management includes crop pest and nutrient management and policy analysis.

Dr. Thomas Warner is completing a six-month appointment as a visiting professional with the W.K. Kellogg Foundation where he participated in the development of a higher education leadership initiative which is linked to another initiative designed to restructure colleges of agriculture and natural resources for the 21st century. Tom also serves as professor and head of the Department of Horticulture, Forestry and Recreation Resources at Kansas State University.

Dr. William M. Welty is professor of management at Pace University's Lubin's Graduate School of Business, director of the Center for Faculty Development and Teacher Effectiveness, and co-director of the Center for Case Studies in Education. He is a recipient of Pace's Kenan Award for Excellence in Teaching. He teaches courses in policy and strategy, public policy, and environmental aspects of management in the MBA program and a college teaching course for the Pace doctoral program. Bill is the author of a number of case studies and monographs on the environmental and ethical aspects of management. For the past several years he has worked on a project to improve teaching at Pace, an activity now supported by grants from the Pforzheimer Foundation and the AT&T Foundation. He has developed a number of teaching workshops based on the case method.

Dr. Ray D. William enjoys both the science and applications of active learning at Oregon State University. He's been involved with creating "farmer-scientist focus sessions" and "discovery sessions" as ways to engage adults in complex decision-making. Recently, faculty facilitators from political science, economics, and horticulture led discussions between farmers, worker groups, and regulators to fundamentally improve regulatory relationships. In classrooms, a diverse faculty have coached group process surrounding natural resource issues while integrating syntheses and analytical thinking using a grab-bag approach and systems diagraming techniques.

APPENDIX A5

Participant Roster

OZZIE ABAYE
Assistant Professor
Crop & Soil Environmental Sciences
Virginia Tech
425 Smyth Hall
Blackburg, VA 24061-0403
PH: 703-231-9737
FAX: 703-231-3075
EML: cotton@mail.vt.edu

FIORELLO ABENES
Professor
Animal Science
Cal Poly University
3801 W Temple Ave
Pomona, CA 91768
PH: 909-869-2089
FAX: 909-625-4537
EML: fjsb11a@prodigy.com

WILLIAM AKEY
Assistant Professor
Plant, Soil & Insect Sciences
University of Wyoming
Box 3354
Laramie, WY 82071-3354
PH: 307-766-5117
FAX: 307-766-5549
EML: wcakey@uwyo.edu

FRANK ALLAIRE
Professor
Animal Science
Ohio State University
2027 Coffey Road
Columbus, OH 43210-1094
PH: 614-292-7142
FAX: 614-292-7116
EML: allaire.1@osu.edu

ANITA NINA AZARENKO
Associate Professor
Horticulture/Pomology
Oregon State University
ALS#4017
Corvallis, OR 97331
PH: 503-737-5457
FAX: 503-737-3479
EML: azarenka@bcc.orst.edu

BARRY J. BARNETT
Assistant Professor
Agricultural Economics
Mississippi State University
Box 9755
Mississippi State, MS 39762
PH: 601-325-0848
FAX: 601-325-8777

ROGER BECKER
Associate Professor
Agronomy & Plant Genetics
University of Minnesota
411 Borlaug Hall
St. Paul, MN 55108
PH: 612-625-5753
FAX: 612-625-1268
EML: becke003@maroon.tc.umn.edu

DAVID BEZDICEK
Professor
Crop & Soil Sciences/Microbiology
Washington State University
Pullman, WA 99164-6420
PH: 509-335-3644
FAX: 509-335-8674
EML: bezdicek@wsuvm1.csc.wsu.edu

ED BRAUN
Professor
Plant Pathology
Iowa State University
Bessey Hall
Ames, IA 50011
PH: 515-294-0951
FAX: 515-294-9420
EML: ebraun@iastate.edu

JOSEF BRODER
Professor
Agric. & Applied Economics
University of Georgia
313 Conner Hall
Athens, GA 30602
PH: 706-542-0751
FAX: 706-542-0739
EML: jbroder@agecon.conner.uga.edu

MARION BRODHAGEN
Graduate Student
Entomology
Univ. of Wisconsin-Madison
102 Henry St.
Mt. Horeb WI 53572
PH: 608-437-4504
EML: brodhagm@bcc.orst.edu

ARDEN CAMPBELL
Professor
Agronomy/Plant Breeding
Iowa State University
1126 Agronomy Hall
Ames, IA 50011
PH: 515-294-7829
FAX: 515-294-8146
EML: acampbel@iastate.edu

DAVID CHANEY
University of California
Sustainable Agricultural Research
    & Extension Program
Davis, CA 95616-8716
PH: 916-754-8551
FAX: 916-754-8550
EML: dechaney@ucdavis.edu

JOHN C. CLAUSEN
Associate Professor
Natural Resources Mgmt. and
    Engineering/Water Research
University of Connecticut
1376 Storrs Rd. U-87
Storrs, CT 06269-4087
PH: 203-486-2840
FAX: 203-486-5408
EML: jclausen@canri.cag

JOE COLLETTI
Associate Professor
Forestry
Iowa State University
243 Bessey Hall
Ames, IA 50011
PH: 515-294-4912
FAX: 515-294-2995
EML: colletti@iastate.edu

JOE CONLIN
Professor
Animal Science, Dairy Extension
University of Minnesota
101 Haecker Hall
St. Paul, MN 55108
PH: 612-624-4995
FAX: 612-625-1283
EML: conli002@maroon.tc.umn.edu

KENT CROOKSTON
Professor
Agronomy
University of Minnesota
411 Borlaug Hall
St. Paul, MN 55108
PH: 612-625-8761
FAX: 612-625-1268
EML: crook002@maroon.tc.umn.edu

DAVID W. DAVIS
Professor
Horticultural Sciences
University of Minnesota
1970 Folwell Ave.
St. Paul, MN 55108
PH: 612-624-9737
FAX: 612-624-4941
EML: davis007@maroon.tc.umn.edu

EDWARD DECKARD
Professor
Plant Sciences
North Dakota State University
Loftsgard Hall
Fargo, ND 58105-5051
PH: 701-231-8139
EML: deckard@plains.nodak.edu

COLETTE DEPHELPS
Outreach Coordinator
Center for Sustainable Agriculture
      and Natural Resources
Washington State University
403 Hulbert Hall
Pullman, WA 99164-6240
PH: 509-335-2887
FAX: 509-335-6751
EML: dephelps@wsu.edu

PAUL DIETMANN
Research Assistant
Center for Integrated Agricultural Systems
Univ. of Wisconsin-Madison
5 Fifth St.
Prairie Du Sac, WI 53578
PH: 608-643-2636
EML: dietmann@students.wisc.edu

MICHAEL DIKEMAN
Professor
Animal Sciences & Industry
Kansas State University
Weber Hall 249
Manhattan, KS 66506
PH: 913-532-1225
FAX: 913-532-7059

CARL R. DILLON
Assistant Professor
Agricultural Economics
University of Arkansas
221 Agriculture Hall
Fayetteville, AR 72701
PH: 501-575-2279
FAX: 501-575-5306
EML: cdillon@comp.uark.edu

TAMMY DUNRUD
Coordinator
Program for Decision Cases
University of Minnesota
411 Borlaug Hall
St. Paul, MN 55108
PH: 612-624-1211
FAX: 612-625-1268
EML: dunru001@gold.tc.umn.edu

JACK R. FENWICK
Assistant Professor
Soil & Crop Sciences
Colorado State University
Plant Science Building
Fort Collins, CO 80523
PH: 303-491-6907
FAX: 303-491-0564
EML: jfenwick@ceres.agsci.colostate.edu

STEVE FORD
Associate Professor
Agricultural Economics
Penn State University
201 Armsby Bldg.
University Park, PA 16802
PH: 814-863-3278
FAX: 814-865-3746
EML: ford@po.aers.psu.edu

EUNICE F. FOSTER
Professor
Crop & Soil Sciences
Michigan State Unversity
160 Plant & Soil Sciences Bldg.
East Lansing, MI 48824-1325
PH: 517-353-1784
FAX: 517-353-5174
EML: fosteref@msu.edu

MARIA GALLO-MEAGHER

Assistant Professor
Agronomy & Plant Genetics
University of Minnesota
411 Borlaug Hall
St. Paul, MN 55108
PH: 612-625-6228
FAX: 612-625-1268
EML: gallo001@maroon.tc.umn.edu

PAUL H. GESSAMAN

Extension Economist
Cooperative Extension
Univ. of Nebraska-Lincoln
205A Filley Hall
Lincoln, NE 68583-0922
PH: 402-472-1748
FAX: 402-472-3460
EML: pgessaman@unl.edu

JOHN G. GRAVEEL

Associate Professor
Agronomy
Purdue University
1150 Lilly Hall of Life Sci.
W. Lafayette, IN 47907-1150
PH: 317-494-8060
FAX: 317-496-1368
EML: jgraveel@dept.agry.purdue.edu

ARTHUR J. GREER

Associate Professor
OSU Agriculture Program
Eastern Oregon State College
Lagrande, OR 97850
PH: 503-962-3613
FAX: 503-962-3444
EML: agreer@eosc.osshe.edu

ORLEN GRUNEWALD

Professor
Agricultural Economics
Kansas State University
Waters Hall
Manhattan, KS 66506-4011
PH: 913-532-4443
FAX: 913-532-5923

ANNE M. HANCHEK

Associate Professor
Horticultural Science
University of Minnesota
1970 Folwell Avenue
St. Paul, MN 55108
PH: 612-624-1706
FAX: 612-624-4941
EML: hanch001@maroon.tc.umn.edu

JEFF HATTEY

Assistant Professor
Soil Science
Oklahoma State University
159 Agriculture Hall
Stillwater, OK 74078-0507
PH: 405-744-9586
FAX: 405-744-5269
EML: jah@soilwater.agr.okstate.edu

C. WILLIAM HEALD

Professor
Dairy and Animal Science
Penn State University
324 Henning Bldg.
University Park, PA 16802
PH: 814-863-3918
FAX: 814-865-7442
EML: bheald@das9.cas.psu.edu

ROBERT C. HERNER

Professor
Horticulture
Michigan State University
A32 Plant and Soil Science Bldg.
East Lansing, MI 48824
PH: 517-355-1857
FAX: 517-353-0890
EML: herner@pilot.msu.edu

ORAN B. HESTERMAN

Professor
Crop & Soil Sciences
Michigan State University
276 Plant & Soil Sciences Bldg.
East Lansing, MI 48824
PH: 517-355-0264
FAX: 517-353-5174
EML: 22626obh@msu.edu

EMILY HOOVER

Associate Professor
Horticulture
University of Minnesota
1970 Folwell Avenue
St. Paul, MN 55108
PH: 612-624-6220
FAX: 612-624-4941
EML: hoove001@maroon.tc.umn.edu

MOHAMMAD JALALUDDIN

Associate Professor
Agriculture/Crop Science
University of Arkansas
Box 4005, 1200 N. University Dr.
Pine Bluff, AR 71604
PH: 501-543-8117
FAX: 501-543-8033

TED JOHNSON

Professor
Microbiology
St. Olaf College
1520 St. Olaf Ave.
Northfield, MN 55057
PH: 507-646-3392
FAX: 507-646-3968
EML: johnsont@stolaf.edu

NICHOLAS JORDAN

Assistant Professor
Agronomy & Plant Genetics
University of Minnesota
1991 Buford Circle
St. Paul, MN 55108
PH: 612-625-3754
EML: jorda020@gold.tc.umn.edu

ROBERT KING

Professor
Applied Economics
University of Minnesota
1994 Buford
St. Paul, MN 55108
PH: 612-625-7028
FAX: 612-625-6245
EML rking@dept.agecon.umn.edu

CHARLES E. KOME

Graduate Student
Crop & Soil Sciences
Michigan State University
Plant & Soil Sciences Bldg.
East Lansing, MI 48824
PH: 517-353-9022
FAX: 517-355-0270
EML: komechar@student.msu.edu

SCOTT KRONBERG

Assistant Professor
Animal and Range Sciences
South Dakota State University
Box 2170
Brookings, SD 57007
PH: 605-688-5412
FAX: 605-688-6170

STEVEN B. LAURSEN

Assistant Dean & Associate Professor
College of Natural Resources
University of Minnesota
235 NRAB
St. Paul, MN 55108
PH: 612-624-9298
FAX: 612-624-8701
EML: slaursen@mercury.forestry.umn

PATRICIA J. LINDSEY

Assistant Professor
Agricultural & Resource Economics
Oregon State University
213 Ballard
Corvallis, OR 97331-3601
PH: 503-737-1416
FAX: 503-737-2563
EML: lindseyp@ccmail.orst.edu

CONRAD LYFORD

Graduate Assistant
Agricultural Economics
Michigan State University
Rm 2 Agriculture Hall
East Lansing, MI 48824-1039
PH: 517-355-1140
FAX: 517-432-1800
EML: lyford@student.msu.edu

DON MACKAY

Research Associate
College of Natural Resources
University of Minnesota
115 Green Hall
St. Paul, MN 55108
PH: 612-626-1205
FAX: 612-625-5212
EML: dmackay@mercury.forestry.umn

CAROL MALLORY-SMITH

Assistant Professor
Weed Science/Crop Soil Sciences
Oregon State University
Crop Science Bldg, Rm 107
Corvallis, OR 97331-3002
PH: 503-737-4715
FAX: 503-737-3407
EML: smith@css.orst.edu

DUANE MANGOLD

Professor
Agricultural & Biosystems Engineering
Iowa State University
210 Davidson Hall
Ames, IA 50011
PH: 515-294-5025
FAX: 515-294-2255

KIM MASON

Student
Natural Resources & Environmental Studies
University of Minnesota
8024 83rd Ave N
Brooklyn Pk, MN 55445
PH: 612-425-1419
EML: chri0177@gold.tc.umn.edu

HAROLD MCNABB, JR.

Plant Pathology & Forestry
Iowa State University
221 Bessey Hall
Ames, IA 50011
PH: 515-294-3120
FAX: 515-294-2995
EML: snabb18@iastate.edu

RICHARD MERONUCK

Professor & Extension Specialist
Plant Pathology
University of Minnesota
1991 Buford Circle
St Paul, MN 55108
PH: 612-625-6290
FAX: 612-625-9728
EML: richm@puccini.crl.umn.edu

DWANE G. MILLER

Department Chair
Crop & Soil Sciences
Washington State Univ
Pullman, WA 99164-6420
PH: 509-335-3471
FAX: 509-335-8674

MARY S. MILLER

Assistant Professor
Agronomy & Soils/Ecology
Auburn University
202 Funchess Hall
Auburn Univ., AL 36849
PH: 334-844-3936
FAX: 334-844-3945
EML: mmiller@ag.auburn.edu

DORIS MOLD

Agricultural Economist
Veterinary Medicine/Clinical Population Science
University of Minnesota
1988 Fitch Ave
St. Paul, MN 55113
PH: 612-626-1277
FAX: 612-625-1210
EML: moldx001@maroon.tc.umn.edu

BERNARD MTONGA

Graduate Student
Crop & Soil Sciences
Michigan State University
1304-H University Village
East Lansing, MI 48823
PH: 517-355-6093
EML: 21922bm@student.msu.edu

HELENE MURRAY
Coordinator
Minnesota Institute for Sustainable Agriculture
University of Minnesota
411 Borlaug Hall
St. Paul, MN 55108
PH: 612-625-0220
FAX: 612-625-1268
EML: murra021@maroon.tc.umn.edu

SALLY NOLL
Professor
Animal Science/Extension
University of Minnesota
1404 Gortner Ave
St. Paul, MN 55108
PH: 612-624-4928
FAX: 612-625-5789
EML: snoll@mes.umn.edu

ARTHUR F. PARKER
Professor
Agricultural Business Mgmt.
Cal Poly University
3801 W Temple Ave
Pomona, CA 91768
PH: 909-869-2208
FAX: 909-869-4454
EML: afparker@csu.pomona.edu

CLAUDIA PARLIAMENT
Associate Professor
Applied Economics
University of Minnesota
218 CLA Office
Minneapolis, MN 55454
PH: 612-625-5733
FAX: 612-625-6245
EML: claudiap@maroon.tc.umn.edu

R. BRENT PEARCE
Professor
Agronomy/Weed Science & Crop Physiology
Iowa State University
1126 Agronomy Hall
Ames, IA 50011
PH: 515-294-3274
FAX: 515-294-8146
EML: rbpearce@iastate.edu

CHRISTOPHER PETERSON
Associate Professor
Agricultural Economics
Michigan State University
2A Agriculture Hall
East Lansing, MI 48824-1039
PH: 517-355-1813
FAX: 517-432-1800
EML: peters17@pilot.msu.edu

GARY PIERZYNSKI
Associate Professor
Agronomy/Soils
Kansas State University
Throckmorton Hall
Manhattan, KS 66506
PH: 913-532-7209
FAX: 913-532-6094
EML: gpiii@ksuvvm.ksu.edu

MARLA REICKS
Assistant Professor
Minnesota Extension Service
University of Minnesota
1334 Eckles Ave
St. Paul, MN 55108
PH: 612-624-4735
FAX: 612-625-5272
EML: mreicks@che2.che.umn.edu

LYNN REINSCHMIEDT
Professor
Agricultural Economics
Mississippi State University
Box 9755
Mississippi State, MS 39762-9755
PH: 601-325-7997
FAX: 601-325-8777
EML: rein@agecon.msstate.edu

SCOTT REUSS
Research Assistant
Agronomy
University of Minnesota
1619 Carl Street #9
St. Paul, MN 55108
PH: 612-625-8759
EML: reuss002@maroon.tc.umn.edu

SUSAN ROZANSKI

Graduate Student
Agricultural Economics
Michigan State University
24 Chittenden Hall
East Lansing, MI 48824
PH: 517-355-9654
FAX: 517-432-1800
EML: rozanski@pilot.msu.edu

ROGER RUAN

Assistant Professor
Biosystems & Agricultural Engineering
University of Minnesota
1390 Eckles Ave
St. Paul, MN 55108
PH: 612-625-1710
FAX: 612-624-3004
EML: rruan@rabbit.bae.umn.edu

RICHARD C. SCHULTZ

Professor
Forestry
Iowa State University
251 Bessey Hall
Ames, IA 50011
PH: 515-294-7602
FAX: 515-294-2995
EML: rschultz@iastate.edu

STEVE R. SIMMONS

Professor
Agronomy & Plant Genetics
University of Minnesota
1991 Buford Circle
St. Paul, MN 55108
PH: 612-625-3763
FAX: 612-625-1268
EML: ssimmons@maroon.tc.umn.edu

ANDY SKIDMORE

Assistant Professor
Animal Science
Michigan State University
121 Anthony Hall
East Lansing, MI 48824
PH: 517-353-9702
FAX: 517-353-1699
EML: skidmore@msu.edu

LOIS BERG STACK

Associate Professor
Applied Ecology & Environmental Science
University of Maine
5722 Deering Hall
Orono, ME 04469-5772
PH: 207-581-2949
FAX: 207-581-2941
EML: lstack@umce.umext.maine.edu

JACK STANG

Associate Professor
Horticulture
Oregon State University
ALS #4017
Corvallis, OR 97330
PH: 503-737-5450
FAX: 503-737-3479
EML: stangj@bcc.orst.edu

SCOTT M. SWINTON

Assistant Professor
Agricultural Economics
Michigan State University
306 Agriculture Hall
East Lansing, MI 48824-1039
PH: 517-353-7218
FAX: 517-432-1800
EML: swintons@pilot.msu.edu

J. ANTONIO TORRES

Associate Professor
Food Process Engineering
Oregon State University
Weigand Hall Room 202
Corvallis, OR 97331-6602
PH: 503-737-4757
FAX: 503-737-1877
EML: torresa@bcc.orst.edu

ROBERT VERTREES

Assistant Professor
School of Natural Resources
Ohio State University
2021 Coffey Road, Room 210
Columbus, OH 43210
PH: 614-292-9792
FAX: 614-292-7432

CHERYL J. WACHENHEIM
Assistant Professor of Agribusiness
Agriculture
Illinois State University
Normal, IL 61790-5020
PH: 309-438-2925
FAX: 309-438-5037
EML: cwachenh@ilstu.edu

RICHARD WALDREN
Professor
Agronomy
University of Nebraska
227 Keim Hall, East Campus
Lincoln, NE 68583-0915
PH: 402-472-1525
FAX: 402-472-7904
EML: agro002@univm.unl.edu

MARK WALLACE
Adjunct Assistant Professor
Natural Resource Science
University of Rhode Island
236 Woodward Hall
Kingston, RI 02881
PH: 401-792-4543
FAX: 401-792-4561
EML: wallam@uriacc.uri.edu

HONG WANG
Graduate Assistant
Agricultural Economics
Michigan State University
Agriculture Hall
East Lansing, MI 48824
PH: 517-353-7898
FAX: 517-432-1800
EML: wanghong@pilot.msu.edu

TOM WARNER
Professor
Horticulture, Forestry & Recreation Resources
W.K. Kellogg Foundation
One Battle Creek Ave. E.
Battle Creek, MI 49017-4058
PH: 616-968-1611
FAX: 616-968-0413

STEVEN WEINBAUM
Professor
Pomology/Tree Biology
Univ of California-Davis
1045 Wickson Hall
Davis, CA 95616-8683
PH: 916-752-0255
FAX: 916-752-8502
EML: saweinbaum@ucdavis.edu

WILLIAM WELTY
Professor
Center for Case Studies in Education
Pace University
861 Bedford Rd.
Pleasantville, NY 10570
PH: 914-773-3879
FAX: 914-773-3878
EML: welty@pacevm.dac.pace.edu

MARY WIEDENHOEFT
Associate Professor of Agronomy
Applied Ecology & Envirnmental Science
University of Maine
5722 Deering Hall
Orono, ME 04469-5722
PH: 207-581-2951
FAX: 207-581-2999
EML: rpt375@maine.maine.edu

WILLIAM F. WILCKE
Associate Professor
Biosystems & Agricultural Engineering
University of Minnesota
1390 Eckles Avenue
St. Paul, MN 55108
PH: 612-625-8205
FAX: 612-624-3005
EML: wwilcke@mes.umn.edu

RAY D. WILLIAM
Extension Horticulturist
Horticulture Weed Management
Oregon State University
ALS #4017
Corvallis, OR 97331
PH: 503-737-5441
FAX: 503-737-3479
EML: williamr@bcc.orst.edu

ALLEN F. WYSOCKI
Graduate Student
Agricultural Economics
Michigan State University
Rm 2A Agriculture Hall
East Lansing, MI 48824-1039
PH: 517-882-3333
FAX: 517-432-1800
EML: wysocki@pilot.msu.edu

HOWARD ZHANG
Assistant Professor
Food Science & Technology
Ohio State University
2121 Fyffe Road
Columbus, OH 43210
PH: 614-688-3644
FAX: 614-292-0218
EML: qzhang@magnus.acs.ohio-state

APPENDIX A6:
Selected References on Decision Cases

Barnes, Louis B., C. Roland Christensen and Abby J. Hansen. *Teaching and the Case Method*, 3rd edition, Boston: Harvard Business School, 1994.

Boehrer, J. and M. Linsky. "Teaching with Cases: Learning to Question," *The Changing Face of College Teaching*, San Francisco: Jossey-Bass, 1990.

Christensen, C. Roland, David A. Garvin, and Ann Sweet, eds. *Education for Judgement: The Artistry of Discussion Leadership*. Boston: Harvard Business School Press, 1991.

Corey, E. Raymond, "Case Method Teaching," *Case Research Journal*, 1982.

Dooley, Arch, and Wickman Skinner. "Case Casemethod Methods," *Academy of Management Review*, April 1977.

Harvard Business School Publications on Case Development and Use. To order copies, call (617) 495-6117 or write the Publishing Division, Harvard Business School, Boston, MA 02163.

  9-598-083   Bonoma, Thomas V. "Hints for Writing Teaching Notes." 15 pp., 1989.

  9-451-005   Cragg, Charles I. "Because Wisdom Can't be Told." 6 pp., 1954.

  9-367-241   Hammond, John S. "Learning by the Case Method." 3 pp., 1976.

  9-584-097   Shapiro, Benson P. "An Introduction to Cases." 3 pp., 1984.

  9-587-052   Shapiro, Benson P. "Hints for Case Writing." 6 pp., 1986.

Hutchings, Patricia. *Using Cases to Improve College Teaching: A Guide to More Reflective Practice.* Washington: American Association of Higher Education, 1993.

Leenders, Michiel R., and James A. Erskine. *Case Research: The Case Writing Process.* 3rd edition, London, Ontario: University of Western Ontario, 1989.

Merriam, Sharan B. *Case Study Research in Education: A Qualitative Approach.* San Francisco: Jossey-Bass, 1991.

Schon, Donald A. *The Reflective Practitioner: How Professionals Think in Action.* New York: Basic Books, 1983.

Stanford, Melvin J., R. Kent Crookston, David W. Davis, and Steve R. Simmons. *Decision Cases for Agriculture.* St. Paul: College of Agriculture, Program for Decision Cases, University of Minnesota, 1992.

Wilkerson, LuAnn and John Boehrer. "Using Cases about Teaching for Faculty Development," *To Improve the Academy*, Vol. 11, 1992.

Yin, Robert K. *Case Study Research: Design and Methods.* 2nd edition, Applied Social Science Research Methods Series Vol. 5. Thousand Oaks, CA: Sage, 1994.

APPENDIX A7:

Suggestions for Maintaining the Momentum

Compiled by Eunice Foster, Michigan State University

The following suggestions for maintaining the momentum developed at the workshop were compiled from the flip chart discussion comments and notecards handed in by participants. The number of participants who made the same suggestion appears in parentheses for the notecard suggestions. If no tally appears, the suggestion was made by only one person.

Listed On Newsprint During the Discussion

1. Electronic bulletin board

   Suggestions were for Internet, Worldwide Web, Gopher, and listserve. Suggestions seemed to indicate that a listserve might be least desirable since it generates a lot of unwanted mail.

2. Refereed cases

   There should be a mechanism for refereed cases, so that institutions will consider them as publications.

3. Clearinghouse

   A clearinghouse is needed for agricultural and natural resource decision cases. The clearinghouse would be responsible for publication and distribution of the cases.

4. Annual handbook

   I don't recall the discussion, but assume the suggestion was for an annual handbook of decision cases.

5. Publish in professional journals

   Authors should submit their decision cases to professional journals such as *HortTechnology* and *The Journal of Natural Resources and Life Sciences Education.*

6. Custom publishing

7. Conferences

   Decision case conferences are valuable and should be held on a regular basis. Suggestions were for every one to three years.

8. Proceedings

   Suggestions were to publish a proceedings for the conferences and to consider publishing on a listserve or on Internet.

Suggestions from Notecards

1. Using cases in distance learning **(2)**

2. Bulletin board/website/listserve **(15)**
   - Networking
   - Get on with listserve
   - Listserve and WWW
   - Could set up a decision case in ag and natural resources website where information and exchanges available on case education. *(Steve Simmons knows of a consultant who would help with this.)*
   - Publish and distribute cases electronically
   - Website for cases, syllabi, etc. plus exchange
   - Second priority would be a website
   - Listserve is not as useful
   - Prefer listserve for updating; don't access web sites very often.

3. Clearinghouse **(17)**
   - Cases needed
   - Easy one-stop shopping for cases
   - Put the greatest emphasis on the clearinghouse function
   - Definitely set up a clearinghouse with some royalty so could build-in longevity. Anything else will eventually fade away.

     Will also build a presence for when private sector moves in.
   - Clearinghouse is essential

4. Production journals **(3)**
   - Some way of publishing cases

- Professional journal
- Refereed journals: *Hort Technology* and *The Journal of Natural Resources and Life Sciences Education* take case studies.

5. Conferences   **(16)**
    - Further workshops, etc.; possibly smaller, regional, to facilitate communication and coordination among those using nearby cases.
    - Regional/national
    - Topics for future conferences:
      * use of cases in natural sciences
      * use of cases in distance learning
      * international cases and application
      * writing and publication of international cases
      * clear description of how to structure a class with decision cases, e.g., discussion with several users of decision cases in classes with their syllabi — what's effective, different directions and why selected
      * consider a session on how to recover from case teaching experiences that "bomb"
      * discuss and show the value of teaching notes; they were a phantom here and many folks seemed concerned
      * posters on new cases
      * work cases on techniques
      * an opportunity for networking
      * more on grading and evaluation of classes using cases; assessment techniques
      * writing case studies
      * more emphasis on teaching techniques
    - Next conference
      * more examples in natural resources
      * more examples of how students are actually involved
    - Conferences should include
      * advanced users
      * how to close
      * logistics
      * relation to different learning styles
    - Conference every three years **(6)**
      * similar structure as this conference
      * another conference on the next phase
    - Conference every two years

- The educational function of this was terrific. Benefits will be greatest if you find ways to attract/facilitate attendance by new people.

6. Download cases through the Internet.

7. Relax the copyright rules or have costs a token amount, perhaps 1¢/case.

8. Encourage professional organizations to include decision cases in their journals, e.g. *Journal of Animal Science.* **(2)**

9. Conduct regional/sub-regional workshop on specific topics; create a case via the workshop.

10. Develop a semi-annual newsletter.

11. What's happening to CASENET? [*Editor's epilogue: CASENET is up and running. To join this list server, send an e-mail request to Tammy Dunrud, University of Minnesota at dunru001@gold.tc.umn.edu. Once on the list server, the CASENET address is: casenet@maroon.tc.umn.edu.*]

12. Copyrighting might stifle the development process. For example, when personal computers first arrived, the main problem was the development of software. Many computer scientists believe that a reason for the rapid development of the computer industry was the sharing of software programs in the 50's to 70's. Protection of cases might stifle our success.

13. Key words should be required for the indexing of all cases.

14. Expand case concept into science and engineering fields.

15. Adapt evaluation using a neural-network-type computer program.

16. Include a contact and address for Kellogg in the participant list.

17. Study and discuss "green sheets" (lists of cases people plan to, or want to, write).

18. Send each participant information on submitting to *Hort Technology* and *Journal of Natural Resource and Life Sciences Education,* and also ask who would like to be a reviewer.

19. Develop a resource of case studies for teachers to acquire cases without much cost.

20. E-mail.

21. Send participants an e-mail of teachers using decision cases.

People Who Want to Participate in Maintaining the Momentum

Bob Herner, Dept. of Horticulture, Michigan State University

Colette DePhelps, Center for Sustaining Agriculture and Natural Resources, Washington State University

Helene Murray, Minnesota Institute for Sustainable Agriculture, University of Minnesota.